Before You Build

AN INSIDER'S INSIGHT INTO THE
DOLLARS AND SENSE
OF BUILDING YOUR HOME

"Opportunity creates a thief"
— Francis Bacon

James Kaufman

Published by *Faux-naïf* House, L.L.C.

Before You Build!
James Kaufman

$14.95
Before You Build
Published by *Faux-naïf* House. L.L.C. www.fauxnaifhouse.com

Library of Congress Identification applied for.
ISBN: 0-9754185-0-5

www.cloverchains.com

TABLE OF CONTENTS

👍 signals a very important issue to consider, "heads-up."

👎 signals unethical or unsavory practices to avoid.

Contents

Contents

Contents

Contents

Dedication

TO

MY WIFE,

LINDA LaCAZE KAUFMAN,

MY COMPANION AND FRIEND,

WITHOUT WHOSE SUPPORT AND

ENCOURAGEMENT THESE PAGES

WOULD NEVER HAVE

TAKEN SHAPE.

ACKNOWLEDGEMENTS

A final word...I would be remiss for not acknowledging the efforts of Kelleigh O. Reynolds and Virginia Gregory, PhD who each in their own way contributed their time and effort to confirm that I had indeed reached substantial completion. Like well experienced construction managers their attention to detail helped sharpen the content to a "final wrap," as in,

"It's a wrap!"

INTRODUCTION

Save yourself time, trouble, and expense by consulting this indispensable field resource to building and remodeling. The guide's easy-to-follow format explores the PROS and CONS of home building and the issues necessary to understand the nuances of the building process. It contains general advice, as well as, practical procedural suggestions for spotting and preventing a variety of unethical building practices.

As a bonus, Part I is devoted to your lot selection, real estate market issues, and financing. The author packs it full of information to help you avoid the pitfalls of purchasing an ill-suited lot in the wrong neighborhood.

Parts I & II are from first hand knowledge, an insider's view, provided by the author, whose first practice as an architect began in 1978. Hence, you will gain constructive insight from his experiences along with examples and a step by step analysis of the building process developed to help protect your interest.

Although, examples compiled from actual experiences and discussions focus on new home construction, they easily translate into multi-family development, commercial construction, or smaller remodeling and interior build-out projects such as condominium units and lofts.

This resource simplifies the building process giving you the knowledge necessary to go forward, implement your decisions, and experience with confidence and satisfaction the completion of your project.

PART I

FINANCE, INITIAL BUDGETING, REAL ESTATE MARKETS AND LOT SELECTION

We were on our way to the attorney's office to sign papers finalizing the purchase of our lot, when we stopped for one last look.

I propped against an eight-foot-high wooden fence on the rear lot line and turned to admire my husband at a distance. He was standing on the front street curb proudly visualizing our dream home to be.

Suddenly, the ground beneath my feet began to move and the wooden fence I was leaning against began to rattle. My ears were suddenly filled with a heavy, thundering, pounding sound that became overwhelming as it drew nearer. The sound filled my body. Finally realizing what it was, I stood there trembling. I dared not turn and look, but out of the corner of my eye, I saw a huge double-decker freight train whizzing by within inches of the other side of the fence.

We had never bothered to look over the fence! I was mortified. I had wished that the ground beneath my feet had swallowed me up.

Chapter 1

GROUND RULES!

(Lots to Consider, Finance, Real Estae Value And Markets)

FINANCING AND ESTABLISHING A BUDGET: HOW MUCH SHOULD I PAY FOR A LOT?

We were rushing to make it before dark. It was time for the final inspection on our new custom home. We had heard horror stories about construction, but we had gotten lucky. We had found this builder who was like a father to us.

Reluctantly, he even sold us a prime lot he was holding in inventory for a relative who just couldn't make up her mind to build. We thought it was a great location far back in the subdivision in a cul-de-sac with no other homes in sight.

With a little finesse, we finally finagled the lot out of him. The only condition was that we pay cash for it before we started construction. He explained to us, he wanted it off the market. We felt a little guilty about prying the lot away from him, but in the end, we felt we had paid a premium for it, so, we figured everyone was even.

Construction had progressed unabated. We were so busy with our own businesses that we only had time to make infrequent visits to the job site.

We wheeled in, parking on the inclined apron of our new driveway. Our home was perched above the street. It had just begun to rain after a long, hot, dry summer. We dashed for the front door, umbrellas in hand.

Before entering, we giggled as we looked around, because, as of yet, there were no other homes on our cul-de-sac. We were alone in our new home for the first time. We were excited!

When we finished our final inspection, which took longer than we had anticipated, it was still raining and now dark outside. I tucked the list of items to be fixed inside my shirt, and we made a dash for the car. We quickly found ourselves in rising ankle-deep water.

The cul-de-sac had turned into a small lake. With no key, we had locked ourselves out of the house. After looking at each other in total dismay, reluctantly, we got into the car. All we could think to do was squeeze each other's hands as we sat speechless.

There was no place we could go, absolutely nothing we could do, until the water in the cul-de-sac subsided. We picked up our feet and placed them on the car seat between us as the water began to cover the floorboards. Neither of us said a word. Early the next morning, before daybreak, the water finally receded. Only then were we able to leave.

It was now obvious to us why the builder had insisted we pay cash for the lot.

Selecting a lot best suited for your needs will take some family self-examination. Your lifestyle will play a defining role in isolating those subdivisions that may complement and nourish it.

If your objective is to build immediately, before you establish a budget, you may want to estimate the maximum amount you can afford to spend on your new home improvements and lot.

Normally, that will depend upon your mortgage loan amount. Your mortgage lender is skilled in estimating the maximum loan amount you will qualify for. Lower interest rates translate into larger loan amounts due to lower payments required.

There are essentially four categories of retail *mortgage lenders* in the marketplace: mortgage bankers, mortgage brokers, typical local bankers, and credit unions.

A mortgage banker, wholesaler, normally differs from your typical local banker where you would establish checking and savings accounts. The main purpose of a mortgage banker is to purchase money in the secondary marketplace and, in turn, sell it to mortgage brokers and bankers for their retail mortgage lending needs. Some mortgage bankers even establish their own retail outlets.

Mortgage bankers may be able to offer you the best *APR*, annual percentage (overall interest) rate, typically having few if any *junk fees* or *points* attached. But, remember, astute businessmen will not under-price a loan. Fees, points, and administrative costs added to the *note rate* determine your APR. Mortgage lenders may saddle the unsuspecting public with junk fees and unnecessary points in order to ramp up their profits.

There are essentially four types of points that concern you; the *origination fee(point)*, the *discount point*, the *lock-fee(point)* and the *premium(point)*. A point or one hundred basis points represents 1% of the total loan amount. It follows that two points represents 2% of the mortgage amount, and one and one-half points represents 1.5% of the total loan amount.

Firstly, an **origination fee** is usually based on 1% of the total mortgage amount, but it can just as easily be based on any percentage (whole or fractional) the lender chooses. Some lenders charge it, some don't.

Secondly, a **discount point** or buy-down fee, usually is a legitimate fee. In order to secure a lower note rate, a lender may ask a customer to pay a discount point up front to effectively buy down the note rate. Usually, this is justifiable, but caution is warranted. This maneuver is sometimes nothing more than a slight-of-hand. We'll explore this in detail below along with premium points.

Thirdly, let's consider the **lock-fee**. To simplify, a mortgage lender charges a lock-fee to hold today's mortgage rate constant. This fee is tied to a future date that you anticipate closing into the mortgage or permanent loan.

Effectively, besides affording you today's interest rate, it ensures that your *ratios* stay within prescribed guidelines so that you remain qualified for your mortgage loan irregardless of market fluctuations. Some lenders even allow your note rate to float downward if market rates go south.

What are ratios? **Ratios** represent your debt to income. With all else equal, lenders calculate your maximum loan amount based on your ratios. Your monthly loan amortization or loan repayment is considered part of your total debt. It follows that rising interest rates will cause your monthly payments to increase. Therefore, a rate increase could effectively disqualify you for the mortgage amount you are anticipating by increasing your monthly payment. This could occur during the construction process, if your mortgage note rate is allowed to float, isn't locked.

Your mortgage lender will advise you about market conditions when considering whether or not to "lock" your note rate or let it "float" with the market. But remember, even the most well intentioned loan officer can't predict the future. Markets are volatile and your rate isn't guaranteed unless you lock. Also, lock-fees hold an interest rate for a specific period of time, which is not indefinite. Four to six months is the norm when considering building a home.

Mortgage lender's policies vary on these timeframes. Determine whether you can complete design and construction within the lock period to benefit from the rate quoted. This time period also becomes important when negotiating your construction contract. You may consider specifying this timeframe in your builder contract, so damages may be assessed if it is missed.

Lastly, **premiums** (points) are paid to mortgage lenders by mortgage bankers for originating the mortgage loan using their money. The amount of premium points paid by the mortgage banker to a mortgage lender is based on the interest rate your mortgage lender chooses for you. Remember your mortgage lender makes his money on premiums paid to him by the mortgage banker, plus fees he charges. The higher the note interest rate above PAR, the more premium points are paid to him by the mortgage banker. Mortgage lenders may offset premium point profits with discount points. They'll charge you a point or more upfront and forego a premium point (or more) hidden on the back end. This allows them to give you a note rate nearer to PAR, maintaining their profit margins, while giving you a competitive note rate. This is the slight of hand alluded to above under discount points.

For simplicity, **PAR** is a note rate with zero points. Charging premium points (hidden in the back end of the note rate) increases the interest rate. Mortgage brokers must disclose premium points charged if the loan is not closed in the wholesaler's name. Mortgage bankers are not obligated to disclose premiums.

When considering your permanent mortgage, discount points and lock fees may well be legitimate, whereas an origination fee may be considered lagniappe.

Junk fees are ancillary administration fees, not absolutely necessary to originate a loan. Legitimate fees may include: underwriting fees, legal fees, title policy premiums, appraisal fees, and funding fees for VA, FHA, and filing fees, etc.

Many other fees may be considered junk fees ultimately unnecessary, functioning as nothing more than a profit center for the lender making the loan. A *processing fee* is usually considered a junk fee. It's up to you to determine which fees may be negotiable. Ask!

By law, in a timely manner (before closing), you must receive a *Good Faith Estimate*, *Truth-in-Lending Statement*, and a *HUD* disclosure (itemized statement for closing) summarizing all costs of the loan; fees and points. Review those documents to determine if your lender is predatory.

You will be required to submit a credit application for pre-approval or pre-qualification. Eventually, when you make or take-down the actual mortgage loan, your lender may require you to put cash (equity) into the project as a percentage of the overall cost, usually 10% to 30%, but 100% loans are available. Equity represents a margin of safety for the lender providing them enough leeway to recover from a soured loan. Whether or not cash is a consideration, you may be required to pay *mortgage insurance* which is another mechanism lenders use to safeguard their investment. In the event of a loan default, mortgage insurance insures the top 20% of the loan-to-value amount, LTV, so the lender may recover.

Once your loan principal is paid down, usually below 80% of the loan to value, your mortgage insurance premium payments should cease. Petition your lender in writing when that day arrives so you may curtail insurance payments which can amount to a substantial amount of your monthly payment.

Cash equity added to the maximum loan amount will be the maximum you will be able to spend on improvements and lot. This is your maximum budget.

A mortgage lender will not normally lend on a home that is contemplated, incomplete or unreasonably maintained. A construction loan is traditionally required to complete construction or repairs before a mortgage can be put into place. Recently, the "one time close" loan was established to eliminate the need for the two loans, the mortgage and the construction loan. Later I will elaborate on the "one time close."

But for now, know that a construction loan is usually based on a bank's *prime rate* plus at least one point.

Therefore, if the prime rate is 9% your construction loan would be 10%.

The phrase "prime rate" can be misleading. Originally, it was established by the largest banks in the nation and represented the lowest note rates available to their best customers, their prime rate. Today, prime rate takes on different meanings. Averaged bank rates printed in the *Wall Street Journal* are considered by many to be the real prime rate, but your lender will establish his own **prime rate** based on his own criteria. Considering the short loan period of a construction loan, the interest rate may float with the prime rate. You may also be charged an origination fee and a few administrative fees to cover the loan origination cost, including inspections, of such a short term. These are essential fees to establishing this loan.

Eventually, in determining your overall budget, you'll want to establish the total <u>cost</u> of the construction loan. That includes the interest to be paid over the term of the construction peiod.

We've already established your maximum budget as your maximum loan amount plus any cash you intend to use. Roughly, figure the construction period to be four to six months. For homes over $500,000 this period may be longer. To establish a rough estimate of the interest to be paid during construction, multiply the amount of your construction loan by the construction loan interest rate and divide that by four. The result should be a good rough estimate of interest to be paid over the six-month life of that construction loan. This simple calculation takes into effect two factors: a construction loan period of six months and the fact that interest will only accrue on money actually paid to your builder and not on money budgeted, but not spent. If you anticipate construction will take longer, the formula will have to be adjusted accordingly.

Add any points and fees charged to figure the total cost of the construction loan. Consult your construction lender for a precise figure.

Your construction lender, normally a local banker, will require a letter from your mortgage lender acknowledging that you have been approved for a mortgage loan that will cover construction costs once your home is complete. You will eliminate some fees and expenses by sharing information between your construction lender and mortgage lender. Hopefully, it's now apparent why qualifying for your mortgage loan comes first.

You're now equipped with all the information necessary to make an intelligent decision concerning financing. A little further into this guide we'll complete the finishing touches by exploring the "one time close." But for now, you're armed with enough knowledge to begin searching for potential lenders offering good rates.

As we progress through this guide, we will explore different methods of establishing building costs. We'll start by establishing preliminary new home costs, which will include design fees, construction loan interest, home improvement costs, lot amenity costs, and lot cost. This you may approach in several ways. We will first establish a market **cost-per-square-foot** to build. By uncovering a cost-per-square-foot to build, we will shape a very preliminary overall improvement budget.

To get started, you may consider consulting local professionals such as lenders, appraisers, architects, builders, and Realtors. They will know with some certainty the average cost-per-square-foot to build. Check with several professionals and compare notes. You may be surprised at the range of prices. This will serve as your first insight into the home building business. There are prices and builders for every pocketbook and subdivision. It still amazes me to find builders pricing the same quality of construction differently depending upon which subdivision they are quoting.

You should inspect the homes of local builders, who construct the style and quality of home you intend to build, and inquire about pricing. Browse your local newspaper for new home advertisements and visit those subdivisions and open houses. Inquire about the prices and quality of construction.

Now, to establish a realistic cost-per-square-foot to build, subtract the lot cost and any additional costs including site amenities (landscaping, drives, etc.) from the overall advertised price. Then, divide the square footage of the home into its asking (remaining) price. I suggest using this method to figure a number of different builder advertised homes. It'll help you establish a good average. This should give you a good overview as to cost-per-square-foot to build.

The point: don't be fooled by square foot cost when comparing prices. Before comparing, determine what is included in the price. For now, **cost-per-square-foot** to build includes the heated living area and not the non-heated garage and porch areas of your house.

Again, when surveying your newspaper, be aware that prices generally include lot cost and lot amenity costs unless otherwise noted. We'll deal with lot cost in a minute.

To determine your home construction cost, do the math. Multiply the average cost-per-square-foot to build by the amount of heated square feet you intend to build. This will temporarily serve as rough improvement costs for budgeting.

Next, take the total amount you intend to spend, your maximum mortgage amount, plus any cash you plan to include, and deduct the home construction cost or improvement costs from that amount. The remainder will be approximately what you can afford to spend on design, porches, a garage, a lot, and lot amenities.

Lot amenities can be extensive and may include expansive driveways, pools, landscaping, etc. Work the estimated cost of the improvements, lot cost, and lot amenity costs against each other to obtain your final budget for each category. It's a good idea to have on hand actual costs (asking prices) of lots in subdivisions you intend to consider. Compare these costs to your lot budget. Are they equivalent? Are you shopping in the right subdivision for your lot?

VALUES, MARKET DYNAMICS, ZONING, DEED RESTRICTIONS: WINNING THROUGH POSITIONING.

When considering particular subdivisions, always take into account the ratio of lot cost to improvement cost. Do not over or under-improve for the subdivision. When considering the optimum ratio of lot cost to improvement cost, there is no golden rule. Though, generally, previous sales in a subdivision or section of a subdivision will establish a ratio of lot cost to improvement cost for that area. Check with your local appraiser to determine lot value.

Lot cost will ultimately be determined by location within the subdivision and community, the quality of life it affords, the market, and the economic dynamics of the area.

At this point, your Realtor is your most valuable asset. To avoid becoming the most expensive or the least expensive home in the subdivision, ask your Realtor to provide you with a list of home sales for the past several years to determine the highs and lows of the market in the subdivisions. Over-improving in a neighborhood can prove disastrous in an emergency, if you must sell quickly. You'll not likely find a willing buyer to pay your price without a reasonable amount of time to market. On the other hand, it's not a good idea to be the least expensive home or under-improve for the neighborhood. You could end up paying a premium for your lot, which means lot cost would be too large based on your budget.

As a general rule, more expensive homes tend to inflate subdivision lot values. As overall lot and improvement values descend in a subdivision, lot values tend to remain artificially inflated due to the more expensive improvements in the subdivision.

Values in a subdivision mimic magnets, each pulling or drawing upon each other. Median-valued homes in a subdivision tend to pull higher-priced homes downward and lower-priced homes upward in value. Higher-priced homes pull lower-priced homes up and vice versa. Generally, this holds true unless a subdivision, as a whole, is moving upward or downward in value. These are rules of thumb that I hope will help you to keep your perspective when considering values in a subdivision.

Convenience to work, play, and school are concerns when considering your primary residence. Ask your Realtor to help you with a list of sales and current lot listings in subdivisions you are considering. Your Realtor will also have insight into the amenities in and surrounding the subdivisions.

Is retail strong in the area, or is it slowly dwindling? Are green spaces within walking distance? Does the subdivision seem cluttered? Is the subdivision well maintained? How are schools rated? Let common sense guide your investigation.

Are values in the subdivision stable? Determine this by examining the age (phase) and economics of the area and the subdivision. The economics are simple to grasp once you understand the basic concepts.

The outermost areas of a city, the suburbs, normally contain the majority of new subdivisions; however, inner-city gated communities are becoming very popular. As you move toward the central business district (CBD), you will encounter mature subdivisions, areas of economics. Still closer to the CBD, some land uses may have changed due to economics, converting for example from single family detached homes arranged on individual lots to denser land use configurations such as multi-family style homes, apartments, or townhouses. You may find other types of subdivisions in the same transitional stage. These may include commercial, retail, industrial, etc.

Land use in unrestricted markets is influenced by land values, amenities, access, and peripheral land uses. It's not uncommon to find certain types (residential, commercial, retail, etc.) of subdivisions converting to other types. For example, inner-city warehouse spaces converting to residential neighborhoods, such as lofts, are presently an inner-city rage.

A city is comprised of hundreds of different types of subdivisions containing their own economics creating a city's dynamics. Whether they are residential or other type, each economic subdivision has boundaries that define its economics, similar to city limits. Generally, larger subdivisions (land area encompassed) tend to promote more overall stable economics.

A subdivision's economics can be affected negatively by non-complementary adjoining land uses or subdivisions. Adjoining land uses at a boundary's edge in negatively impacted circumstances usually become transitional in nature with values less stable or in decline.

To simplify, boundaries define areas of economics, small or large. They can be visual, such as a railroad track, or as seamless and understated as perceived property values. They can be defined simply by views, a line-of-sight, a change in deed restrictions from subdivision to subdivision, a value (sales price) change among subdivisions, lot sizes, traffic count, or something as visual as a fence, stand of trees, park, drainage ditch, street, railroad, or actual land use change, for example, a detached single family subdivision adjoining a multi-family land use. Any of the foregoing may represent lines of demarcation or economic boundaries.

Establishing boundaries is important in your search for stable subdivisions because they define the economics of the neighborhood (subdivision). An area of economics can be as small as a city block, a neighborhood, a subdivision, or as large as a city, state, or nation.

Each area, within its boundaries, contains its own economic dynamics. These are independent and interdependent of each other.

Remember, just because you are within what appears to be a stable subdivision doesn't mean there aren't economic forces acting upon it.

Theoretically, in cities like Houston where there is no zoning, land use is determined by the free market. Control of land use in these circumstances is sometimes established by **deed restrictions**, while in the majority of other cities, land use is controlled by zoning ordinances. **Zoned** and **non-zoned** cities each have their PROS and CONS.

The biggest fear, real or imagined, is usually regarding adjoining land uses at boundaries edge. What will go up next door? But, normally, there are winners and losers in both zoned and non-zoned markets where subdivision fringes take the brunt of any economic challenge.

Economically speaking, non-zoned free markets

are better at determining appropriate land use, the highest and best use, compared to land use dictated by arbitrary land-use formulas: zoning. The worst economic impact of non-zoned markets seems to be insufficient utility services to serve unbridled real estate development, which usually becomes a developer's problem to resolve. Although the same problem surfaces all too often in zoned communities, becoming even more pronounced mainly due to the lack of sustained growth and development due to land use restrictions imposed, which translates into lack of funding for utilities.

Zoning ordinances establish *quasi-monopolies* often controlled by politics. Zoning artificially inflates or devalues property, which in turn trickles down affecting other economic dynamics. For instance, in zoned communities, rents may be artificially inflated due to the lack of additional available land to meet the needs of apartment-unit demand. With zoning, land use is established, good or bad. That is until a proposed use change becomes politically charged. Then, anything can happen.

Zoning traps real property into artificial land-use, and property values can suffer. As economic dynamics evolve dictating land-use change, many properties experiencing transitional forces lie in waste becoming economic burdens to their owners and communities due to land use restrictions. Politics, more often than not, control land-use change in zoned communities.

Ideally, whether zoned or non-zoned, you are searching for subdivisions with a history of strong sales. Strong sales confirm strong demand and hence strong values. What are strong sales? Small gaps between asking and sale prices may be considered strong sales and good indications of advancing property values. Sale prices gapping consistently lower than asking prices may be an indication of falling values.

Quick sales or short marketing periods are usually good indications of strong values. Prices will be edging upward or at least stable. In a few markets where demand is exceptionally strong, price wars occur, resulting in the ultimate price paid for a house being well over the original asking price.

For each subdivision you are considering, ask your Realtor to research and compare the number of *days on the market to sell.* Compare subdivisions to determine the strength of sales in the overall area. The shorter the days on the market to sell the stronger the market!

Subdivisions and houses in the upper tiers of value have their own inherent problems. As prices reach the high bar in a community or subdivision, there are fewer and fewer prospective buyers who can afford to pay the higher prices. Therefore, days on the market may increase considerably when analyzing high-end markets. Compare apples to apples when it comes to residential neighborhoods or any real estate.

It may be helpful to imagine a market as a triangle with one side resting flat representing its base. The base represents the lower-priced homes while the highest point, the apex, represents the highest price a buyer is willing to pay for a house in that market.

The area within the triangle represents the buying public. As you ascend toward the apex, the available buying public diminishes. As you descend toward the base, the lower-priced homes, the quantity of the buying public increases. It might be useful to remember this concept when evaluating prices within a subdivision.

Another factor to consider is the life cycle of the subdivision. For example, transitional subdivisions with eroding values, decreasing sales prices, exacerbate trends toward land use change, i.e., older subdivisions with larger lots transitioning through a stage where lots are being subdivided to accommodate smaller homes of lesser value. More about this below!

To assess the overall economics, ask lots of questions. Query appraisers in the area to learn more. They should have a good pulse on the market because they are constantly confirming sales information and analyzing trends. Just make *certain* the appraiser you select is an expert in your area, actively working the market of interest and has comparables in your immediate area to back up his opinion. Remember, economic subdivisions. Comparables one street away in the wrong direction could make a huge difference in value. Most appraisers work with their head in the sand. I'm personally familiar with an appriaser brougth in from across town to appraise a lot in an area unfamiliar to him. He appraised it for $30,000. It was worth at least $300,000. My home sat next to it. I knew the area well. As a safeguard, I'd recommend having your Realtor give you a list of comparables for the area, but make sure she knows the area. I can't overstress this.

Check zoning ordinances for change. If there is no zoning, refer to the subdivision deed restrictions. The absence of zoning and deed restrictions are sure signs that non-conforming uses will create unstable values.

If your subdivision has deed restrictions, realize that they may not be perpetual. Determine the period of time the restrictions will remain effective, as well as, what provisions are provided for their extension. Do the deed restrictions automatically renew? Review and examine the legal mechanism provided that ensures reinstatement. Will the deed restrictions remain in effect through your ownership period, through the next purchaser's ownership, and well beyond? If there is doubt about time remaining, consult a professional, an attorney, or title company professional for an opinion.

Strong deed restrictions can translate into strong values, except in cases where land use in the subdivision or surrounding neighborhoods is in transition. Dwindling sale's prices may start to predominate. In this case, the economics of the area are changing and are forcing land-use change. The transition might not be toward the land use you desire.

When examining subdivision covenants and deed restrictions, look for restrictions that promise strong values. These may include an established and active architectural review committee, an established minimum square footage of improvements, and an established minimum allowable lot size. Restricted use of exterior materials can also dictate conformity, therefore, potentially strong values.

Other signs that may indicate solid values include: older homes being razed to accommodate new construction, extensive remodeling of existing homes, well maintained neighborhoods, large lots, younger couples moving in, strong homeowner associations, strong subdivision covenants, well enforced zoning ordinances, conformity of housing design, strong sales, and vigilance over non-complementary land-use infringements by surrounding properties.

NEW, MATURE, TRANSITIONAL LIFE CYCLES: FROM A DISTANCE, IT ALL LOOKS GREAT!

Further value assessment may be achieved by considering the economic stage or phase of the subdivision. There are essentially three stages to the **economic life cycle of a subdivision**: new, mature and transitional.

New subdivisions or the earliest stages of a

development may afford the best potential for quick value increase, but they often tend to be the riskiest phase of the subdivision-development life cycle. Value dips may occur due to poor planning, poor (loose) subdivision covenants, poor quality of construction, slow absorption, or emerging incompatible land use surrounding the periphery of the subdivision.

When purchasing in the early stages of subdivision development, false price ceilings should be considered. If you're anticipating a move shortly, beware of false price ceilings created by unsophisticated builders holding prices in check on homes they build repetitively. This affects values in the subdivision, artificially keeping values low.

Appraisers must use recently sold properties in the neighborhood for comparables to determine the value of your home. Often, stagnant new house pricing thwarts chances of value increase.

The mature phase of a subdivision is its most stable economic stage. Prices reach their peak potential and remain stable until the next stage, transition, occurs.

The last phase of a subdivision's life cycle is the transitional stage. Land-use changes begin to occur in this phase. Land use outside a subdivision may or may not be complementary. Values may erode due to economic pressures within the neighborhood itself or economic pressures created by land use peripheral to it.

Effectively, at this stage, a new economic subdivision's dynamics are evolving, with new economic boundaries and life cycles. This new subdivision will evolve through the three development phases or life cycles until economic pressures call for another land-use change. As you might imagine, each phase may last only a few years or perhaps decades or centuries depending on the economics and the circumstances within and surrounding the subdivision.

Once you have isolated a particular subdivision, visit it on weekends and after work. Isolate potential lots and do some visual snooping.

Is there ample off-street parking, or do the streets become clogged with parked cars after hours? Again, let common sense guide you to the correct conclusion.

Generally, the most valuable lots will be located well within the subdivision far and away from such annoyances as major intersections, railroads, community pools, and other noise, visual, and stink intrusions. Smells from public-waste water facilities in the surrounding area brought by changing winds can also be detrimental to values.

Visit the subdivision during a hard rain because tremendous land development over the past two decades has created major water-shed problems. Drainage systems are overburdened due to poor planning and under capitalization.

With all else considered, a few mature trees are an asset. They clean the air while providing cooling and soil stability. On the other extreme, heavy overhanging foliage may promote mold and mildew, creating a wealth of health concerns for certain individuals.

Also, make sure trees are not infringing into the *building envelope*, especially if there are *tree ordinances* in your area. Ordinances may prevent their removal or trimming. Be informed; national registration of trees restricts their removal.

AND THE WINNER IS…WHAT'S IN A BUILDING ENVELOPE?

Visualize the **building envelope** as an ice cube sitting on your lot that has melted into a particular shape due to outside influences. The remaining shape represents the total maximum allowable cubic airspace within which you may legally build your home. Its shape is derived from deducting yard setback requirements, easements (power line, drainage and other easements that may reach high into the sky or deep into the ground) and building-height restrictions (including codes, covenants and utility easements) from the overall lot area and airspace above, sky. The resulting form we shall call the building envelope. This three-dimensional area defines the cubic space within which you are allowed to build.

STOCK YARDS, SCHOOL YARDS, SIDEYARDS: WHAT'S IN A YARD?

Yard setbacks are areas of your lot that must remain unobstructed. They occur adjacent to and parallel to the perimeter property lines (boundary lines) and their size (width) will be defined in building codes, covenants, restrictions, and zoning ordinances.

LOTS TO THINK ABOUT: CORNERS, INSIDERS AND CULS, WHAT'S YOUR PREFERENCE?

Our new home on our dream lot was finally finished. The house ended up costing us significantly more than first estimated. It had become a struggle taking more of our time than we had anticipated. We spent endless hours arguing with the builder, finally deciding we'd never go through that again.

About a week after the job was complete, when we had settled in, a representative from the homeowner's association dropped by. He asked if we were aware that the subdivision covenants required a corner-lot owner to install a sidewalk the length of the lot along the side street. We were shocked and dumbfounded as he explained the subdivision covenants to us.

We explained to him that we weren't aware of that requirement, and the builder who sold us the lot and built our house hadn't informed us either. We had contracted the builder to take care of everything necessary to comply with all the codes and restrictions.

The representative shook his head in dismay as he advised us that the sidewalk had to be constructed immediately or the homeowner's association would levy a daily fine against our property until it was completed. We were aghast when he informed us that they could even place a lien against our property until the fine was paid off. They could even sell our home!

We had just eked by with the house financing, barely qualifying for the new loan. All the added cost had eaten up our cash reserve, and we had just financed a house full of new furniture, which we intended to pay off over the next five years. We were strapped for cash and didn't know where to turn.

We'd have to ask our banker if he would extend us enough credit to put in a two-hundred-and-seventy-five-foot concrete sidewalk. But what would it cost, and who would we call to do the work?

Generally, when considering typical subdivision lots, there are three types: corner, interior and cul-de-sac.

Corner lots are the most accessible and usually larger. They also demand higher prices. Trade-offs include increased street noise and larger and more severe yard setbacks, such as a substantial side-yard setback. Construction can prove to be more expensive due to enormous street frontage. The front-yard setback requirement, as well as, a large side-yard (street) setback will impinge the potential "building envelope" far greater than an interior lot's setbacks.

On the other hand, *interior lots,* although not as readily accessible, are generally less expensive and smaller. It is arguable that interior lots are less private due to adjacent neighbors. Some maintain that interior lots are more secure when considering children.

The third is the *cul-de-sac* lot. Usually, less street noise is associated with cul-de-sac lots and those on the approach to a cul-de-sac. Delivery trucks and other through traffic avoid the dead-end aspects of cul-de-sac streets. Resulting pie-shaped lot configurations can create difficult building envelopes.

Floor plan adjustments must be made to accommodate cul-de-sac lot shapes due to limited front street access. The cul-de-sac lot value will normally be higher in price than an interior lot of the same square footage.

Of course, final lot selection will always remain an individual choice based on market availability, lifestyle, and financial considerations.

LOTS: MORE TO THINK ABOUT, OTHER ISSUES WORTH CONSIDERING.

We left for our lake home in Austin after negotiating a contract with a builder to build our dream house. He had an old set of plans that seemed to meet our needs, and we marked them up with changes while he took notes. He assured us he would have new plans drawn with all the changes. He told us he would take care of everything.

Upon returning, we met some friends at the construction site to inspect the progress. We all nodded in agreement that it was coming along nicely until my wife noticed that the room sizes weren't at all what we had discussed, and the windows we had planned to overlook the water at the rear of the house weren't in the right locations.

Immediately, I called the builder who informed me that he had instructed the person staking out the house to cut it down (the size of the house) because the home he had promised us wouldn't fit on the lot.

I wasn't quite sure what to do or say. He had never called us about the changes. I couldn't believe it. He told me he had the newly revised plans but wouldn't produce them for us to see. He informed me it was my word against his. This guy had totally hoodwinked us.

We had looked long and hard for this lot before making the final decision to purchase it. Now this!

The walls were made of concrete. What could we do?

There are many decisions to be made in the early stages of planning that should not be left to chance.

Correct *orientation* toward the sun, a more subtle issue, is a benefit you can enjoy over the life of your home. It can make a huge difference in energy-consumption cost savings, as well as, affecting interior natural lighting exposures. Cooling glass rooms with western exposures increases not only utility consumption, but also initial cooling equipment size requirements, capital expenditures.

Will the *shape of the lot* accommodate your planned improvements? Pool area and future expansions must be considered.

To quickly assess a lot, grab a copy of the survey and find the scale of the drawing. What we're about to outline doesn't have to be perfect. We're just trying to determine if the lot is large enough to accommodate the proposed improvements.

To start, draw lines parallel to the property lines indicating yard setbacks and easements: front, side, and rear. Call the local building-official's office (permitting department) if you need help with yard setback requirements. Also, review your subdivision deed restrictions. A recent survey should locate any utility easements and encroachments. Work around any easements to determine how they may affect the buildable shape, the building envelope. Don't forget to place your pool on the lot if that's a consideration. Inquire about any related ordinances.

Structure encroachments into easements are

forbidden. You may be able to work out a compromise with owners of overhead easements by agreeing that any damage to your improvements caused by their access is your responsibility. If you can live with that, they may possibly allow you to build ancillary improvements within the easement, but get it in writing.

Remember, you are developing a building envelope as described above. After all side yards and easements are drawn onto the survey, the remaining cubic space within your lot is a good representation of your allowable buildable area. Most subdivisions allow two-story houses, but check the height restrictions for the lot by consulting your covenants, homeowner's association and local building codes.

Is your remaining lot shape, the maximum building envelope, large enough to accommodate your planned improvements, the square footage and height of your home, and other lot amenities you desire?

Other factors that may influence the size and shape of your improvements will include additional green space requirements, which may be defined in subdivision covenants or zoning ordinances. Minimum or maximum allowable buildable square footage will also be defined in the subdivision covenants.

Exactness, leave that to the design professionals.

Consider including a contingency in any lot purchase agreement that will subject the purchase to a satisfactory review of the building envelope constraints by your home designer.

He will review all governing restrictions, which may include homeowner association rules and by-laws, your recorded plat, city or county planning commission ordinances, deed restrictions and restrictive covenants. He will also communicate with the building official about matters that are unclear.

His research and analysis may uncover further restrictions that must be considered in your design including: square-footage requirements, height restrictions, material usage, colors, roof pitch, landscaping requirements, fees, utility availability, demolition, and other special restrictions, including historical areas and moratoriums on construction.

Also, have him review *wetland* issues, which may affect your lot usage resulting in an undesirable building envelope.

Remember, in your original construction contract have your builder acknowledge that he has visited your lot and has become familiar with all aspects surrounding your site including the scope of construction and any and all local codes, restrictions, and covenants. This will serve as a further check in verifying that design parameters used were correct.

Drainage and areas prone to storms (hurricanes) are other major considerations. Drainage issues may dictate that your home be elevated above natural grade. Flood insurance may be required, creating a life-long expense, not to mention the catastrophic experience, financial and emotional, of a flood or storm.

Land elevation and the contours of the lot are factors that influence the final costs of improvements. Fill dirt is expensive to haul, compact, and landscape.

Stand on the curb and observe the street and lot characteristics. Will water flow naturally away from your lot during heavy rains? Walk the length of the lot to determine whether it sits lower or higher than adjoining properties. Does it contain low spots, swales? Beware, laws prohibit lot elevation changes that divert storm water onto adjoining properties. If your lot grade must be raised, extensive drainage systems may be required to carry storm water away from your property.

If your lot is low, a sump pump may be required to lift and disperse storm and waste water (sewage). This will necessitate life-long pump maintenance costs, extra-utility expenses, capital expenditures, and inconvenience when the pump becomes clogged or defective.

Are curbs and gutters complete, or is the storm drainage served by open ditches? Insects breed in warm weather, incubating in standing water. Are the streets and underground storm systems in place? If not, when are they projected to be completed? Get it in writing!

Are there sufficient utilities, kind and quantity to satisfy your proposed improvements? If you intend to use gas appliances, is gas utility available? If sanitary waste systems are not public, you may need a soil percolation test to determine whether the size of your lot is sufficient to meet health requirements and accommodate the placement of waste water field lines.

AT THE TOP OF YOUR GAME, OR IN THE CAT BIRD SEAT.

It was early Friday evening when we submitted our full price contract with no contingencies to purchase a lot. We were building our dream home. We didn't want anything in the contract to get in our way. The lot was perfect, and we couldn't figure out how we had missed it before. We were tickled with it. We had lost other lots by not acting fast enough.

Our real estate agent informed us that the selling agent wanted the earnest money deposited with her when we presented our contract. I gladly wrote the check for $10,000.

After several hours, our agent called and informed us that she had finally located the seller's agent and had delivered the contract and earnest money, but due to unforeseen circumstances, the seller himself had been called out of town. He was to return Monday.

Late Saturday afternoon, we decided to drive out to the lot and dream about the possibilities. We were leaning our butts against the car hood, feeling all smug and surveying the lot, when, just over the horizon, we spotted a silhouette of what appeared to be a large wide-bodied commercial airliner. Before we could scramble inside the car, it thundered overhead lowering its landing gear as it released some type of gray fluid that spewed all over our clothes and car.

We were awestruck. Our purchase offer had no contingencies. It was for cash. Had we just bought the farm? The seller's agent worked part-time, which would make her exceedingly difficult to locate.

I knew enough about contracts to realize that if we didn't rescind our offer before the seller accepted it, we were screwed. At the least, we would lose our earnest money if the seller didn't sue us for specific performance. How could we have been so stupid? We had to hurry! There was only a few hours left before the seller's return.

It can't be emphasized enough: give yourself ample time to investigate, uncover, and understand all the issues surrounding a lot purchase. Two weeks to thirty days is standard, but request more time if necessary. This is your **due diligence** period.

In your lot-purchase contract, make the purchase contingent on your satisfactory findings. It allows you time for research and discovery before your final decision to purchase, transferring title into your name. I insert a statement into the *special provisions* section of the contract similar to this. "Purchaser is hereby granted a period of thirty days from the *escrow date* of this contract to complete due diligence. Within this period, purchaser, at his sole discretion, may elect to terminate this contract in writing rendering it null and void with earnest money returned to purchaser in full, promptly at the time of notification."

Financing is another contingency consideration. But

if the lot is owner financed, a financing contingency may have no place.

☞I use language similar to, "This contract is contingent upon purchser securing financing, satisfactory to the purchaser at his sole discretion, within forty-five days from the escrow date of this contract. If financing is not forthcoming with no approval contingency within this time period, this contract automatically becomes null and void, and all earnest money shall be promptly returned to purchaser."

The next item of importance is what I call a **fuse.** This limits the time the seller has to accept, counter, or reject your offer. It automatically terminates your offer without having to submit a withdrawal letter if it is not accepted within the timeframe specified. It also creates urgency for the seller to consider your offer and helps to eliminate the possibility of the seller *shopping* your contract to other potential purchasers.

☞Again, under special provisions of the contract I use wording similiar to, "this offer is null and void if not accepted by seller and delivered to purchaser by 5:00 p.m. tomorrow (the drop dead time and date).

The **drop-dead** time and date can be any date and time, but it must give the seller a reasonable timeframe in which to act.

Sometimes, multiple contracts are parleyed in order to drive up the price. If you think there are other contracts working, make an offer, have it delivered, but don't sign it. This eliminates any chance of getting entangled in something of no benefit to you. The fuse along with an absent signature, puts a little more urgency on the seller to act. He'll probably realize he's dealing with a sophisticated purchaser. Sign only after the seller has agreed to all terms acknowledged by his signature on the contract. But care must be taken, this tactic can backfire and offend the seller causing him to refuse to counter. Use it with caution.

👎 I never relinquish *escrow money* until the contract is consummated. If possible, always place escrow money with a disinterested third party, an **escrow agent**. Always insist that the escrow be placed with a title company or an attorney who will close the sale. This eliminates a potential priority conflict. First and foremost, escrow money to be refunded is easier to obtain from a third party. By using a disinterested third party, the client aspect is removed, and the intent of the contract becomes the main focus.

👍 Have the escrow agent acknowledge and date the contract upon receipt before returning a copy to you! That date becomes the **escrow date** mentioned above upon which all contingencies are tied.

👍 If somehow as a purchaser you become responsible for paying part of the sales commission on the lot, make sure to eliminate any wording in the contract related to real estate commissions being earned at consummation of the contract instead of at **closing**, the day title is transferred to you. Normally, the seller pays the real estate commission. Question anything else. Thoroughly review your contract and understand every aspect of it before you sign.

Seek help if you don't understand terms. Never assume anything. Keep a copy for your files of every stage of the negotiation process along with a fully signed final draft, the executed contract with a copy of your acknowledged security deposit.

Before closing, you may elect to purchase *title insurance*. **Title insurance** is issued in two parts: owner's and lender's policies. Lenders will insist on their coverage, but you have a choice. Ask your closing attorney specifics about what you are really buying when you purchase title.

PART II

DESIGN AND THE BUILDING PROCESS

As far as my builder is concerned, he's a crook; he's a thief, and he's a liar. I've written the Attorney General bringing this to his attention. He informed me that there wasn't much he could do. I've written the board of contractors twice, and they won't do anything about it.

Unfortunately, this gives him (the builder) the reins to lie and steal and do anything he wants. To think, he used to be president of the local homebuilder association.

I do know for a fact that there are a number of people for whom he has built houses who are very upset with him. I remember one couple who had been living in their house for six months and the inside still had not been finished.

In our case, he was running a year beyond the scheduled completion date, which forced us into temporary housing. Finally, we did get permission from the local building official to store some of our personal belongings in the house. We had the locks changed. The house was finally secured even though it remained unfinished.

He's deceitful, he's dishonest and he's hurt a lot of people, not only financially but also emotionally.

Chapter 2

INSIDER RULES!

(Knowing The Game Puts You In The Cat-bird Seat)

"THE SKINNY" OR THE REAL DEAL!

...this guy was born to lie. We had contracted for a hurricane home with the new concrete-wall system. The builder told us the house would resist winds up to 200 miles per hour.

Well, during construction, I started noticing some irregularities. The plans called for our roof to be built with 2 x 8 rafters 16 inches on center, but what we got were 2 x 6's 24 inches on center. Our plans clearly showed that the rafters shouldn't be cut or notched. There was total disregard for that.

After he built it, he told us, "I only build to minimum code." When we were in his office and asked him to build us a hurricane-resistant home, he knew what we were talking about then, and now, all of a sudden, we're getting a house built to minimum codes. When he was putting his pen to paper, he billed us for a hurricane house, but we didn't get it.

I learned one thing from all this, our builder was a Con-artist. Choose your builder v e r y carefully!

Logic seems to confirm that house plans and specifications permitted for construction would dictate how a house is to be built. This is not necessarily so if left to the local building official or code-enforcement officer to officiate.

Hear this! Many governmental officials, code-enforcement officials, who issue building permits based on plans submitted for construction, take the clear stance that they are not responsible for what the builder constructs. Their only obligation, as they see it, is to make sure that **minimum code standards** are met. Say that again! That *minimum* code standards are met, as set out in the codes adopted by the city or county!

Therefore, if your plans call for a more elaborate roof than codes require, the building official won't obligate himself to insist that the builder install it. That responsibility is up to you or your representative. There is some logic to what appears to be madness. Over the years, code officials have found that owners change their minds as often as day rolls into night. Code officials have also found it wiser to stay out of the fracas and not become referees. Also, the code-enforcement officer will not necessarily place himself in a position of verifying products used for construction.

By sticking to minimum standards, the code-enforcement official meets his obligations of enforcing the codes.

THE PLAYERS: WHAT PART DO YOU PLAY IN ALL THIS?

...it's likely you won't get a second chance;
so, you better get it right the first time...

Each player in the construction process has his own agenda. The main players, the owner(s), the design professional, and the builder together with the help of peripheral players, such as the lender and building official, orchestrate the building process. What follows is a thumbnail profile from each prospective.

The owner is usually a partnership between two people. For simplicity, let's divide the roles into male and female. I've witnessed this partnership disintegrate time and time again, especially under the pressure of construction. If your marriage or union isn't stable, a common misconception is to think that the construction process will bring you closer. The opposite will happen! Many marriages can't take the pressure of construction. Examine your motives! Construction is a process that takes on a life of its own. It's not over till it's over.

The following capsule provides a simplistic view of the owner's participation. It is a teaser for further self-examination and role playing.

The **male role**, competitive in nature, is expecting fair play with a level playing field. A top priority is the budget. He expects to get what he has contracted and paid for. Normally, his approach is game-like: "You win a few and lose a few." The male also expects to be able to handle construction problems because he equates this as being part of the male M.O. More often than not, he fails to realize that construction is a very sophisticated, confusing, and technical business, and that the construction documents, a precursor to construction, will soon become the major focal point in the process. Their focus and scrutiny will grow in intensity based on how well the job progresses. Friendly relations between owners and builders can disintegrate into brawls. The owner must take ample time to ensure that his construction documents are in order before construction begins. We will define construction documents in a minute!

A word of caution! Unscrupulous builders will literally obliterate the male ego due to the fact that the male isn't educated, experienced, or trained in the mechanics of the construction process or construction documents. Therefore, the first step is to make sure your construction documents are complete (whole), and you are familiar not only with the construction process but with the paper work necessary to proceed.

Early on, simple mistruths or information from your builder that just doesn't quite jive may be initial warning signals. Your senses are keen during negotiations. Don't disregard confusing statements. During early negotiations, you may learn more about your builder's integrity than you care to know. Pay attention!

Eventually, finding that you've been duped into engaging Mr. Slick Eddie, quick decisive action may be necessary to rid yourself of his insanity. Believe me, it's never too late. Remember, the builder is the master of construction, knowing all the angles, practicing them daily. Alone, the average homeowner won't be able to handle an unscrupulous builder. It's nothing to anguish over. Seasoned professionals often can't handle fast-talking unscrupulous "Eddies". Cut your losses and move on.

To remedy, consider engaging a competent construction representative and get back on track, or discharge your builder and contract anew.

Care must be taken in selecting your representative. It's a tough role that must be filled by the correct person. Someone who is confident about himself and what he knows. He should be formally educated, trained, and experienced in construction. Being astute to the *good ole boy's club* can pay off. If your first representative doesn't get the job done, replace him. Remember, it's not you.

Today, if a person can pick up a hammer (nail gun), he calls himself a builder. *Now is not the time to be timid.* I assure you, by the end of the job, you will be glad you acted. It will save you time, money, and heartache.

I highly recommend engaging a competent construction representative from the onset of the project. Seek his advice when writing the construction contract. Do it! It's cheap insurance. It will save you money!

When drafting your construction contract, make sure to define "you" as meaning *you or any of your representatives as may be designated from time to time.*

Now, let's consider the **female role**. She is motivated by socio-economic status and also has concerns about cost. Traditionally, she has not been as competitive as the male. She is a mediator, a communicator. She has a good sense of right and wrong and also wants a fair deal and is usually not as confrontational as the male. In a confrontational environment, I've found her more willing to compromise, put things behind her, and move on.

Her concerns about the well-being of her family often take precedence and weigh significantly in her ultimate decision process.

An unscrupulous builder may also target her. He may hold both owners hostage by letting the job sit idle until they scream UNCLE. Then, he'll get what he wants, perhaps the final payment!

After a delusional experience of making that payment without the job being completed, male egos usually tend to echo similar sentiments. For example,

I ended it because of my wife. She had had enough. She literally forced me to "close". The builder was running six months over the construction deadline. We had sold our other house, and our furniture was on the truck. We had no place else to go. What could we do, but move in?

At closing, I presented the builder with a laundry list of items that needed to be addressed, but once he received the final payment, we've never seen or heard from him again. Someone on his staff just keeps giving us telephone lip service.

Females on the other hand seem to resonate another message.

I had to end it before my husband had a heart attack. I could see the builder wasn't going to budge. What were we going to do? Anyway, the builder assured me not to worry. I can't help it if the man doesn't tell the truth.

Examine your role and your partners and strategize to determine what approach you will use in the months ahead to handle job site issues. You don't want to become overwhelmed and lose control.

Another player, the **design professional** sells *time* in the form of plans and specifications. Usually, his livelihood demands that he make as much money in the shortest possible amount of time. That's not all bad, if in fact, he produces a complete set of plans and specifications defining your ideas along with quality construction standards. The faster the better? Here the owner must press the design professional to make sure the plans and specifications are complete eliminating the need for changes during construction.

Change means time. Time is money. Therefore, someone will have to pay. The liberating factor here is that you are not the expert in this field either. Before engaging his or her services, ask as many questions as necessary to achieve the comfort level you desire.

NOW is not a time to be bashful. He's the expert; take advantage of his knowledge. Get a sense of what he is all about. Is he compatible with your personality and your design ideas? Just because he designs homes, doesn't mean he will be capable of fulfilling your design needs.

To complete the construction process successfully, you'll need to enhance your level of knowledge. Spend quality time with your design professional. Take nothing for granted. Make sure you fully understand your plans and specifications.

Make sure your design professional knows cost of construction. If he relies too heavily on the builder to supply cost, he may be putting you at a disadvantage. To be effective, the design professional must know costs.

If two-dimensional plans are difficult for you to interpret, have a three-dimensional model made. Find a talented high-school or college student who can help you. Models don't have to be constructed through your designer, and they don't have to be expensive. Anyone who understands scale can help. Possibly someone taking a drafting course at a local trade school or high school would be available to help you.

At the end of the design process, the design professional will present you with final plans and specifications. By then, you must feel very comfortable knowing your way around the set of plans.

Warning, if you can't interpret your plans, your builder will be glad to interpret them for you, the way he "sees" them.

The other player is the **builder.** *He is selling his expertise, his financial stability, his ability to finish your house according to plans and specifications within a defined budget and prescribed timeframe.* His ultimate agenda is to profit from his service, and his plate is full of opportunities.

Again, before starting construction, complete your plans and specifications, demand a firm construction price, a firm construction deadline, and state them all in your contract, a contract closing the loopholes. Together, the combination of these will serve as the road map you'll follow over the next several months. Being thorough, before construction commences, will ensure a successful outcome.

In your original construction contract, you must define damages incurred by you due to construction delays or time overruns. One of the few exceptions for not completing construction on time would be days of rain, game rained out. But after the roof is in place, rain days are usually no longer excusable.

Also, define allowable *mark-up* (builder profits and overhead, a percentage of the cost) for authorized substituted materials and also allowable mark-up on labor for authorized additional work. Have the builder attach a list of hourly wages for each trade or job and reference the list in your contract. These costs should be at his fingertips. Have the builder attach a list of *unit prices.* **Unit price** refers to a quantity such as a single shrub, a square foot of paint, a square foot of tile, etc. They represent the total cost including costs to install such items as labor and materials at the job site. These prices cover everything from the square-foot price to install ceramic tile to the square-foot price to paint. You will use these unit prices to add or deduct from the construction cost during the job depending on changes elected.

As a general crosscheck, consult your **schedule of values**, a cost breakdown by job category your builder will supply to your lender as a checklist for payment. Compare the numbers. Do the sum of unit prices and schedule of values agree?

National standards are also available for cross-referencing costs. Check your local bookstore. Means Estimating Handbook by R. S. Means breaks down time requirements, crew sizes, and crew mixes necessary to complete specific construction tasks, including: hanging doors and windows, painting, etc. Make sure you choose the correct guide.

Your builder's cost may be lower than cost derived from national standards, unless you are in a geographically isolated area. One reason, your builder may use "scab" labor whenever possible, therefore lowering his costs. Have your builder attach a list of wages or fees for each trade or job and reference that list in your contract. This will aid in determining adds and deducts, credits and debits. Never contract for work on an open-ended hourly basis, "time & material". Always demand a firm price and finish date. The CON is never to finish, effectively gouging you.

When completing your budget, include the cost of landscaping, fencing, lawn sprinklers, sidewalks, etc. Make sure these items are included in your construction contract.

Your builder will compute his labor cost using the following "units" for pricing, question anything else:

Tree demolition - by the tree
Landscaping – by the item or square yard for sod
Fencing – job or linear foot of fence plus gates, etc.
Earthwork - cut and fill of soil by the cubic yard
Slab forms - square feet of slab area
Termite treatment - square footage of slab
Slab form - square footage of slab
Slab setup - cost of vapor barrier and steel
Slab pour - square footage of slab
Slab finish - square footage of slab
Framing - square footage of house plus extra for arches and specialty items. Normally, includes black-in, completely enclosed house with building wrap, windows dry-in.
Electrical - square footage of house, extra for floor plugs and specialty items. Price may include fixture hanging, phone lines, doorbell and appliance installation and fixtures. Verify!
Plumbing - square footage of house, fixture count and pipe. May include gas fixture and appliance installation and fixtures.
Sheetrock - square footage of sheetrock material, extras for heights over ten feet and other specialty features including such items as arched windows.
Painting - square footage of house or square footage of

wall-surface area and type of finish.
Trim Work - square footage of house, includes setting doors and hardware and baseboards. Extra for setting base and upper cabinets (charge by the cabinet unit).
Carpet - by the square yard
Tile - by the square foot
Countertops - by the linear foot of counter top
Roof - by the square (100 square feet)
Driveways - square footage of surface area

Talk to subcontractors and vendors to determine how they derive their costs. For instance, consider the category earthwork. Earthwork is figured by the cubic yard of soil that is either brought to the site or taken away. To calculate the amounts correctly, you must be aware of a few factors. Soil expands when it is cut (dug up) and must be compacted when it is placed (filled). To fill one cubic yard hole requires more than one cubic yard of soil because of compaction. Different compensating factors must be used for different soil types. Don't let this frighten you. I've chosen earthwork because it is the most complex and difficult to estimate of all the categories.

When considering the category of sheetrock, it is customary to figure labor cost based on the amount (square feet) of whole sheetrock pieces brought to the job site and not the square footage of sheetrock hung in your house. There is a difference. Check your dumpster.

Never be defensive about asking for prices. There are no secrets when it comes to costs. You can gather them yourself if inclined. It's logical and rational to determine all costs before construction commences. You want to make sure your builder has included everything in his pricing. If he hasn't, he'll cut something to make up the difference to recapture his profit. If he's an honest professional, he should want to share pricing with you. He's a businessman trying to make a fair profit for a fair day's work. Isn't he?

You're not the expert in this field, and your builder must have reasonable answers. 👎 A good answer is not, "I don't have time or you won't have to worry about that." If there is a hidden agenda, now is the time to uncover it.

👎 If he is dodging you now, how much time will he have to spend with you once construction commences? This is not a game of who can outwit whom. The CON is to make you feel inadequate about cost.

Remember, builders and subcontractors are not rocket scientists and many have very little education, even though I've found attorneys turning to home building just to make a living. They come with their own agenda, but many builders come from other disciplines they left behind. It's a testament to how easy construction is to enter.

Contract ratification is critical for the builder. It's the most important stage of the process for him. He must convince you to sign a contract, after which, you may never see him again. Therefore, nail down prices before execution of the contract, or there is little hope afterwards.

If you elect not to hire a full-time representative, at least consider hiring one on a consulting basis. It's money well spent. I urge you, <u>at a minimum,</u> to engage a competent construction specialist to help you through the *contract stage* to ensure that you document in writing all the information necessary to establish a good working relationship. The contract creates the basis on which to build and make informed decisions.

👎 I hate to destroy illusions, but an attorney with little or no actual experience or education in the construction process will be of little help to you at this stage. There are attorneys proficient in the issues we discuss, but state bar associations I've contacted have made it virtually impossible to locate such experienced attorneys, if by nothing more than simply refusing to give reference as to those who are experienced.

The majority of attorneys I have interviewed seemed to have little or no working knowledge or experience of the construction process. I also understand that many states consider construction a specialized area of law. A brilliantly written "empty" contract isn't worth the paper it's written on. One thing the public has yet to understand is that an attorney is a technician, not necessarily a businessman. There is a vast difference. If he's not a construction expert, all he can do is rely on the builder to inform him. As a warm up, I'd suggest buying him a copy of this information source. Then couple him with a construction professional.

Choose your construction representative with the care of choosing a doctor. I can't stress that enough. Engage someone who is not only experienced in the construction process but is also educated in the construction process. There can be a huge difference in the knowledge base between the two.

Remember, the builder negotiates and argues on a daily basis with some of the most hardheaded and stubborn people on this earth, subcontractors.

Let that be your warning. The builder is accustomed to negotiating in a hostile environment. A little crying and yelling during construction on your part will most likely fall on deaf ears.

Chapter 3

GAME PLANS FOR EVERYONE!

(Plans And Specifications)

The job had been shut down for three months while we waited. Our builder had constructed a roof pitch lower than allowed by the subdivision covenants. He had assured us before starting construction that he had reviewed the plans and made sure they were in compliance with all the codes and regulations. We trusted him! It had now become evident that he was more interested in getting us on contract and into our pockets than in taking care of our best interest.

Our house was a concrete-walled house. The subdivision developer told us we had only two choices. Unless every homeowner in the subdivision would agree in writing to let us proceed, we could either walk away leaving it as a concrete monument to the subdivision or bulldoze it, which we couldn't bear to think of. We already had over $100,000 invested in the construction, not including the cost of the lot.

IF A PICTURE IS WORTH A THOUSAND WORDS, WHY DO YOU NEED A CONTRACT? LEARNING HOW TO STACK THE DECK IN YOUR FAVOR.

Plans, specifications, and your construction contract are the guidelines, the mix of information necessary to complete your project on time and within budget. Together, the three play independent yet integral roles in defining the scope and cost of construction. They are the essential elements of the construction documents.

In your construction contract, be sure to define **construction documents** as a combination of plans, specifications, the construction contract, anything each references, change orders, field orders and all governing codes and regulations. In your construction contract, state that construction shall comply with ALL the construction documents and also set forth the priority of documents. In other words, state which document takes precedence when there is conflicting information.

Plans, working drawings or construction documents, which can be synonymous, include the site plan, floor-plan layout, elevations (vertical views of walls, interior and exterior), and schedules (door, window, hardware, electrical, plumbing, etc.) that define the relationship of spaces, rooms. They indicate the quantity of materials and items, their shapes, and the locations to be installed within the proposed improvements. Quantities of specific door types; their sizes, profiles, and in which rooms specific doors are to be located are an example. The overall floor plan shape is defined along with wall sections and details of construction. Site plans locate your home improvements and site amenities.

You will find schedules, a list of specific items in the plans including doors, door hardware, windows, appliances, plumbing fixtures, etc. More information can be found about these items in the specifications.

Specifications work with the plans and typically address quality of materials, methods, and procedures of construction, which help to further define construction techniques; therefore, quality and costs. Quality can be defined by simply specifying an item made by a particular manufacturer, to be installed per the manufacturer's recommendations. That's all it takes. The item has been identified along with its manufacturer who correctly defines how it is to be installed, according to their recommendations.

Quality may also consist of defining a construction procedure. For example, it may describe a method or procedure to follow for the construction of a pool or a roof system.

Referencing an industry standard such as ASTM (American Society of Testing Materials) may also define and describe the quality and method of installation.

These are just a few scenarios used in specifications that help define the quality of material and/or method of installation.

If you like throwing away money, here's your chance. Ignore the specifications. Although they are written in plain English, specifications can be dry and boring to the typical individual not involved in construction, but they are essential to the successful building process. Together with the plans, they define the quality and quantity of construction; therefore, cost. *Lack of specifications, essentially, leaves the quality of construction up to your builder.*

Take the time to examine every item and procedure that has been specified for your home, including: the quality of paint (manufacturer and grade), required surface preparation before painting, how the paint is to be applied, the thickness of the paint coat specified, number of coats, key holes, doors, door hardware, windows, kitchen and bath cabinets, counter tops, lighting fixtures, bathroom tile and fixtures to water heaters, etc., making sure that no interpretation is left to the imagination, leaving nothing to chance.

Caution: if you need further information about the items specified or "called out" on the plans, stop before consummating your contract for construction and visit the vendors from whom your materials will be purchased. Ask the builder to supply you with a list of those vendors. Don't hesitate to take pictures of the items to be purchased for your home or ask for a *cut sheet* on the items. A **cut sheet** is published by the manufacturer or distributor and contains not only a picture of the item, but also a description highlighting important features and installation procedures to be followed. If they don't have a cut sheet, try the internet. Look for the manufacturer and select the item.

Attach each cut sheet and picture to the contract; reference them in the body of your contract, and make sure to keep a copy for yourself in a safe but convenient location.

If something is not specified, literally described in your construction documents, chances are, when it comes time to choose, the builder will signal that your choice is more expensive than what he has budgeted; therefore, construction costs will increase.

Starting construction without a complete set of plans and specifications with all the items selected will come as a natural instinct. It's an exciting time. You've been dreaming about it. You can't wait to get started. You can't believe your dreams are finally going to take shape, but now is the time to use self-control. Slowly, put on the brakes. Take a deep breath. A few more days tying everything down will ensure your success.

Only fools rush into construction without:

1.) *complete plans*
2.) *complete specifications*
3.) *total construction cost*
4.) *a deadline for construction completion*
5.) *and a tight thorough contract (loopholes closed)*

Together with the plans and specifications, the **construction contract** is an essential piece of the construction puzzle. It is as important as the plans and specifications. It is the glue that holds it all together.

Seek advise from professionals when negotiating your construction contract. Caution: your rights are up for grabs. Be steadfast in your resolve to set the stage for commencement of the construction process.

SELECTING A GAME PLAN: WHERE CAN YOU PURCHASE A SET OF PLANS?

It is imperative to understand that plans and specifications serve separate functions, and your construction documents must include plans, specifications, and your construction contract to be complete. Insist that plans and specifications be completed before signing the construction contract. Never allow the preparation of your plans and specifications to become part of your construction contract. Don't do it. They should be prepared by a third party independent of the builder.

There are many sources for plans including architects, home designers and off-the-shelf plans which include plan rooms, plan books, and in-house builder sources. Recognize that plans are only as good as the person producing them. Involve yourself thoroughly in their preparation to make sure your plans represent your needs. Tie everything down. Rid yourself of loose ends. Before signing the contract for construction, take time to make revisions and define materials and methods of installation, not during the construction period.

Now, let's examine design professionals. **Architects** are educated, experienced and trained professionals in the art of design and construction. Like individual tastes, there are architects who specialize in different types of design. States require continuing-education courses as a condition of licensing and also establish a code of ethics to which each architect must adhere. On the average, architects are well trained in design and construction documentation (plans and specifications) preparation and have a good sense of the steps required to ensure quality. They, like most others, gather their pricing from the marketplace. They are trained to interpret client's needs in the form of plans. They tend to use A.I.A. (American Institute of Architects) contract forms.

Many of these forms are perplexing to the general public. Over the years, due to the litigious society we now live in, there has been a concerted effort by the A.I.A. to ensure that all parties to these contracts are treated equally and fairly. Some have voiced the opinion that owners may have been better served by earlier versions.

When using A.I.A. documents, make sure to read them thoroughly. They are in plain English, and you must understand what you are reading. If not, seek help! There are several types of construction contracts, (large or small jobs, abbreviated or full-blown contracts) offered by the A.I.A. Make sure you choose the one right for your job. Include a set of "general conditions" (separate cover) in your reading, often neglected, but referenced by most A.I.A. contracts. They are an essential part of A.I.A. contracts. The contract addresses the specifics of a particular project, while the General Conditions are less specific to a project, more global in nature.

Note, some states mandate that an architect is the only person qualified to create construction documents. Others require that an architect be used only if the square footage of the house exceeds a certain predetermined size, say 4,000 square feet. Otherwise, plans for improvements not meeting or exceeding the defined minimum square footage can be prepared by anyone. Your local building official at your city permit department will guide you there.

At this point you might be wondering how architects charge for services. Most architects charge one of three ways: a flat fee, by the hour or a fee based on a percentage of the construction cost. Methods and amounts may be negotiable.

I advise you not to use the same architect who prepares your plans to represent you on site. The code of ethics may get in his way.

I personally would want someone looking out for my interest with no fiduciary responsibility toward the builder. Standard A.I.A. contracts may gray that line. Architects serve as great representatives to interpret the construction documents and make site inspections, but hire them independently of those preparing your construction documents. You don't want their liability for construction-document contents clouding their opinions.

Another source for construction documents may be a **home designer** or **designer**. Dedicated designers don't necessarily have formal education, and many states don't require that they be licensed; therefore, continuing education is not a requirement. They tend to focus on the homebuilding industry and may well be experts in their field of practice because of their focused experience. They may be lacking in the area of structural concepts, specification writing, and quality of materials, but the same goes for architects. Designers are normally less expensive than architects because designers carry less risk and tend to have no licensing or continuing education requirements. Designers usually charge a flat fee or invoice by the hour. Designers in my opinion wouldn't make good representatives for site inspection.

Another source for plans is **plan rooms**. Plan rooms can be found in major cities around the country. For $2,500 and up, you can select from very detailed plans that have been stockpiled over the years. Many of these plans have been revised often to satisfy customer needs. They're also updated to reflect the latest construction methods and code requirements. Changes to suit your needs are readily available at a reasonable fee.

Plan books by comparison can be found at any local building supply or local book store. They fill a niche, even if only to heighten your senses, reminding you of your likes or dislikes. Plans from these sources can be obtained for a nominal fee, usually under $600, and most providers will allow a local designer or architect to make revisions to suit customer's needs. Providers encourage that plans be upgraded to meet minimum code requirements in the local area of use. These plans are usually adequate for permitting, but leave much to the imagination.

All plans are copyrighted and should not be used or copied without authorization.

Now let's discuss the quality of plans. There are generally two types of plans, a *full set* and a *builder set.* The latter by far is the most widely used in the homebuilding industry. The providers outlined above may prepare either type. At first glance, the average person probably won't be able to distinguish the differences.

The **full set** has more detail, and therefore is more expensive to produce. Architects produce these sets of documents, but good designers are also capable. These sets are usually a "one shot," complete with foundation details, and casework (millwork or cabinet) plans detailed and developed for a particular client along with accompanying specifications. When using a full set, your builder will have to pay particular attention to detail; therefore, construction cost may be higher if your builder is not accustomed to the quality of work specified. Materials, methods of construction, installation, and finishing methods are outlined in the full set. This ensures that nothing is left up to the builder to personally select or choose as he sees fit.

The **builder set** of plans on the other hand will contain the same type floor plan and elevations, wall views, but will usually lack many details, casework plans, and specifications. This set is adequate for permitting but leaves much to the imagination—the builder's imagination!

Putting it into perspective, **with a full set, you get what you pay for if you pay attention; with a builder set, you pay for what your get.** Between the two, there can be a significant difference in quality, but not necessarily in cost. You may be charged for construction quality that you don't receive if not defined in your plans and specifications.

Review your plans thoroughly and insist that anything not defined in the plans and specifications be outlined and described in detail and included as an addendum to the contract for construction. The contract ties it all together answering such questions as what, when, how, how long, and how much you will pay for construction. Without a thorough contract, you are at risk.

It's like taking a cross-country trip without a map. You know where you are and where you want to go. You've got pictures of where you're going, but you don't have the slightest idea of the paths to get there. Always reference your plans and specifications as an exhibit in the contract by date issued, design professional, and title.

Reduce to writing everything that you expect, want, and are willing to pay for. Take nothing for granted. Nothing compares to the importance of this. The builder may take it upon himself to select everything that is not specified, described, spelled out, or referenced.

Sometimes, it's not enough just to see a profile drawing of an item when it really needs to be *called out*, described. Sometimes the price of an item varies tremendously. Take for instance, crown molding. Have your design professional call out the specific shapes, which I assure you have all been well documented for his convenience in standard catalogues. It will save you time and money.

Furthermore, it is not a good idea to simply specify a brick exterior when you can select the brick before the contract is signed. Physically, make a brick and mortar selection to confirm the price. Have your builder "lay-up" different brick selections and patterns for you to make an informed selection. If you don't make your brick selection before construction begins, have the selection process outlined in your specifications. Pre-select brick and mortar even though you haven't yet decided on the combination. It's not cost-effective to change your mind once the brick is in place on your home.

While we're on the subject, brick prices and labor costs vary tremendously. Brick prices are based on size, color, quality, and production location. Color and quality usually determine mortar cost. Brick masons charge according to size and weight of brick and by the thousands laid with decorative work like columns and decorative coursing adding extra costs. Brick masons don't all charge the same rate.

Be cautious of a builder who insists that construction start without having complete construction documents. It's a prescription for disaster. Unscrupulous builders push unsuspecting owners into commencing construction with only an estimated cost, a loose set of plans, and no definite completion date.

I've witnessed the final cost on an "estimated" $300,000 home settle in at $400,000 plus at completion. Serious price adjustments made during construction indicate poor planning, unethical construction practices, or both. In this case, it was both. By any stretch of the imagination, a $100,000 *increase* is extremely difficult for any owner with a budget to absorb considering the original estimate.

Consider your bank's position. Even if your

banker is willing to increase your loan limit to cover such an increase, they will ordinarily have to update the original appraisal to justify the cost increase. The cost might not be supportable as value in your neighborhood, which simply means that you, the owner, would have to absorb those additional costs with cash out of pocket. This may affect not only your construction loan limit, but also your permanent loan (your ratios), your mortgage. Caution is the recommendation here.

You've probably chosen your builder based on a recent home you have seen and admired. Be extremely cautious if you've elected to start without complete plans and specifications. I strongly don't recommend what I'm about to suggest, but as a last resort to protect your interests when specs and complete plans are not possible, reference in your contract a home built by your builder.

Better yet, reference a *model home.* By doing this you have something to support claims which will ultimately arise.

Be exceedingly careful when looking at a model home. Make sure what you are admiring is what you intend to build. Don't take the builder's word for it. Spell it out in your contract. Reference the finish work, including but not limited to the cabinets, counter tops, trim work, exterior and interior finishes, and reference the address of the model home as defining the standard of quality and finish work you expect in your home. A bit tricky, but as a last resort, it's better than nothing. Take pictures, attach copies and reference them in the body of your contract and keep an executed copy for yourself. For instance, if the model home has oversized kitchen countertops make sure you get a picture and reference that in your contract.

A last word of caution, once construction starts and you start making changes, you lose.

It's that simple!

Chapter 4

KNOWING HOW TO STACK THE DECK!

(Initial Costs vs. Maintenance And Operating Costs)

...I called the builder once I noticed the vinyl siding was defective around the lower edge of the house. I couldn't imagine how I had missed it during the walkthrough. I guess at the time I was more interested in the interior finish work than anything else.

I was a nervous wreck until he arrived. Well, about a week after I called, he finally made an appearance with the vinyl subcontractor in tow. After a quick inspection, he informed me he wouldn't fix it. He told me that whoever had mowed and edged my lawn had destroyed the vinyl siding with their weedeater.

It wasn't his problem, he said, but I explained to him that no one had cautioned me about the vinyl siding and yard equipment. Shaking his head and smiling, they climbed into his Lexus and drove off.

BEATING THE ODDS: PAY NOW OR PAY LATER!

When selecting material and equipment, it is important to compare the initial cost including installation costs to future maintenance and operating costs. Don't forget, maintenance and operating costs span the life of your ownership.

As mentioned earlier, one of the first considerations should be *lot orientation* in relation to the movement of the sun. Before purchasing the lot, study your proposed improvement orientation to make sure you can take optimum advantage of the sunlight and its warmth.

Simple orientation adjustments may compensate for harsh westerly sun. Over the life of your home, sun orientation could amount to substantial savings in operating costs. Orient your home to take advantage of afternoon and morning sunlight.

Other considerations may include room volume, cubic feet, and the amount of exposed glass directed toward sunlight. As room volume increases, air conditioning and heating costs also escalate. *High ceilings,* besides being more expensive to frame and finish, ultimately demand larger cooling equipment and more energy to heat and cool.

Configuration of spaces can also play a significant part in cost savings. As an example, square shapes require less material to construct than do rectangular shapes.

Take, for instance, the square shape of sixteen square feet. The square can be enclosed with four equal sides, each side four feet in length for a total of sixteen linear feet of exterior wall space.

Take the same square footage, but instead enclose it as a rectangular shape. Calculate two sides as two feet in length and two sides as eight feet in length for a total linear exterior footage of twenty feet. That's four feet of extra wall space to envelop the same square footage and that's a small space! Sums are exponential when considering a house. Now, you're beginning to get some idea of why one house with 2,400 square feet under roof may cost more than another with the same square footage. Linear wall space makes a significant difference in the ultimate cost.

Open floor plans with less linear wall space won't necessarily cost less. Large roof spans, beefier structurally to accommodate the open floor plan, may negate any perceived cost savings. Also, open floor plans are more challenging and expensive to air condition and heat.

Glass houses are expensive to build, upkeep, cool, heat and maintain.

Costs of materials are always subject to much debate

when weighing selections. Tradeoffs between initial costs of materials and maintenance costs create the ebb and flow of budgeting.

Traditional *wood framing* is usually the least expensive and most practical to use, due partly to the overabundance of trained labor required to erect it. Wood is usually readily available; therefore, lowest in cost. Wood material prices, however, are subject to fluctuation based on market conditions including tariffs, and market demand and supply; hence, prices may skyrocket. *Metal framing,* on the other hand, may be an alternative if wood prices have escalated. Normally, metal framing, a little more expensive than wood, is also slightly more expensive to erect and large roof spans can become a design and expense issue.

When using metal, special thermal breaks between the exterior sheathing and the metal studs should be implemented to achieve the same insulation qualities as wood. Over the life of the home, lack of termite infestation and lower insurance premiums may offset the initial cost of metal framing. A composite structure using the best of each, metal and wood, is another consideration.

I.C.F. (insulated concrete forms) construction, a relatively new concept in home building, represents the high end of the cost spectrum. I.C.F. units are hollow polystyrene blocks stacked atop each other and then filled with steel reinforcing and concrete. I.C.F. construction offers the possibility of a stronger wall, reinforced concrete with excellent insulation qualities and the added feature of sound attenuation. One of the biggest drawbacks of the I.C.F. unit is termite infestation. The polystyrene blocks which remain in place for insulation purposes present the termite with excellent tunneling opportunities, up through the polystyrene, into the wooden roof plates and rafters above!

Remember, no matter what argument the builder makes, never let I.C.F. units come in contact with the soil, not even insect-treated soil. I suggest they be at least eight inches above any finished grade. Make sure this is clear in your plans.

In coastal states, people are being hoodwinked into using I.C.F. construction for hurricane-resistant purposes. You should think about the whole structure, not only the walls, when considering hurricane-resistant issues. Usually, a builder will construct the roof the old-fashion way, typically using wood framing. If this is true in your case, the overall construction is no better than the weakest part, the roof. Why pay for a 200-mile-an-hour-resistant wall when your roof is still susceptible to blowing off under much less wind pressure? It's the first thing to go, and unless otherwise specified and built to comply, the builder will only build the roof to meet minimum code requirements. As outlined elsewhere in this reference, minimum strength requirements specified by code are much less than the strength claims made by I.C.F. manufacturers.

Another hidden factor that may affect the strength of I.C.F. construction is shoddy placement of concrete within the I.C.F. units, especially in house construction where labor is often unskilled. Improper concrete mix or lack of proper vibration when pouring will result in honeycombed concrete instead of the solid mass needed to fill the I.C.F. units to obtain the desired design strength.

If you may be subject to water surge from a hurricane, I for one wouldn't want to return to find only concrete walls standing, when all else has been washed away. I assure you, I've experienced the surge affects first hand. Your insurance company may deduct their value from compensation due you and in the end you will probably still have to remove the walls and start over. Not a good idea!

Trade-offs between the initial cost to install exterior finish material and the required maintenance of that material after initial installation, especially when considering exterior finishes, deserve much deliberation. For instance, the initial cost of *brick* makes it one of the most expensive exterior finish materials available to install, but it remains relatively maintenance-free throughout the life of the home; therefore, it is well worth considering compared to some of the new high-tech exterior coatings that may demand endless maintenance and could even render your home unlivable. The new synthetic stuccos may fall into this latter category.

Stone, conventional stucco, and materials such as *Hardy Plank* (cementitious material formed to resemble wood siding), *wood, metal* and *vinyl* siding are other possible exterior finishes. I recommend caution when considering new high-tech materials similar to *E.I.F.S.* (exterior insulation finishing systems), a synthetic stucco and its derivatives. Major problems are evolving due to moisture penetration and mold and mildew growth within the walls behind such materials, and much litigation is brewing. Most problems seem to occur due to improper installation. But an attorney well entrenched in this debate on behalf of some homeowners enduring these affects, once confided in me, "It's not a question of if, it's a question of when you'll have problems with synthetic stuccos."

Vinyl and metal siding materials fall under the category of relatively inexpensive and maintenance-free materials. Make sure you are familiar with all their nuances. A weedeater, lawn equipment, will devour vinyl. If vinyl is your choice, make sure your designer specifies a skirt surrounding the lower perimeter of your home made from a material that will hold up to lawn-care equipment. Wind can rain havoc on either of these sidings. Also, metal siding is susceptible to denting and oil canning, buckling in extreme heat. Hardy Plank siding, mentioned above, may also be susceptible to wind damage and sag.

For color change, you may paint any of those materials, while the manufacturer of Hardy Plank recommends painting it for longer life. In fact, you can order Hardy Plank pre-finished. Wood, either cedar or redwood, is more expensive initially to install, and maintenance is a life long consideration. Wood comes in many configurations from shingle style, to tongue-and-groove applications. It's a good rule to stay with the higher grades of a wood product when considering wood for exterior siding. I always stick to translucent stains when using expensive wood siding, which allows you to enjoy its texture and warmth while providing adequate protection. Paint hides wood qualities and is costly to maintain.

Today, most interior stud walls are covered, or hung in sheetrock. Sheetrock, gypsum board, and drywall all refer to the same product, gypsum sandwiched between two layers of paper. It forms the basis of what will eventually be your finished walls.

Gypsum board finish is important when considering the ultimate finish material. Whether you intend to finish it with paint, flat or stippled, wall covering or leave it bare, I recommend contacting the Gypsum Association for sheetrock preparation options. They publish industry-recommended standards to assist specifiers, owners, and contractors.

One of their most informative publications comes in the form of a video, demonstrating methods of preparing gypsum board to receive different types of finishes. The paper publication is entitled Levels of Gypsum Board Finish, GA-214. Your plans and and specs should include their recommendations.

Chapter 5

WIZARDS ALL AROUND US: BUY THE BEST AND NEVER HAVE TO WORRY ABOUT THE REST!

(Know What You Are Buying)

We just had to have it. We both hated the feel of cold ceramic floors during the winter. I found this heating system in a reputable home improvement magazine. I called, and the manufacturer informed me that they didn't have a distributor in our area, but they consented to sell it to us direct. They told me anyone could install it if they just followed the instructions. So, I ordered it. I had to threaten the builder to get him to install it. He charged us dearly for that installation.

It never worked. It never heated the ceramic tile as advertised. The manufacturer wouldn't give me any relief because they claimed it had been installed improperly by an unauthorized contractor. The builder was no help. He claimed that he had followed the installation instructions. It wasn't his problem. I ended up with over $7,000 of worthless wiring in the bathroom floors under the ceramic tile. It never worked and still doesn't work. It was a tough lesson, but I learned it well. If I ever have it to do again, I'll force the builder to buy and install any extras I want, making them his responsibility.

TRADE-OFFS: IT'S ALL ABOUT CHOICES!

Generally, I avoid materials new to the market that seem too good to be true. Home builders are usually more apt to use these new materials. Some even purchase franchises and push the products without disclosing their ownership interest!

I.C.F. units are a good example. But anything including stamped concrete and kitchen cabinets could fall into this category.

Other materials that warrant further investigation are some of the new paper thin "woodish" flooring materials. I would suggest conferring with several wood floor vendors and installers offering traditional ¾" oak floors. Query them about flooring materials. Then question several representatives offering the new technology in wood flooring systems.

Installation cost represents the majority of the cost when considering wood floors. You'll find that there isn't much cost savings between the two. Any rational person can determine, if they just listen, which floor will give them lasting value with less maintenance. Since repairing a damaged floor is a major consideration, flesh out the truth concerning this aspect of maintenance.

Listen carefully! *Vendors will tell you what they don't want you to know.* It's the human psyche. Our litigious society cautions them to expose the truth by filtering it into the conversation. And don't expect an unbiased opinion from someone who just delivered 2,000 square feet of ¾" oak wood flooring.

Other high-tech items that are puzzling include the new paper-thin kitchen cabinet finishes. Bringing these products to the marketplace had to be a real feat. Kitchen cabinets, like floors, take severe abuse. Ever noticed the corners of a cabinet door peeling off like paper or countertops turning green? That's the backing.

Beware of Mexican brick. Nothing matches its warm texture and soft colors. What makes it enduring, its softness and porous texture, also renders it a poor value. These bricks eventually crumble into piles of clay. Again, it's a material that looks great new, but has no lasting value. Many codes have now outlawed its use.

Stay with *name brands* when considering appliances

and equipment, those with good warranties. Strongly consider *high efficiency ratings* when choosing a heating and cooling system. Check with your local power company for available *rebates* when considering high efficiency equipment; otherwise, someone else may pocket them.

Use *extra insulation* where possible. It's inexpensive and will last for the life of your home. *Blown insulation* is best for attic spaces, but not for wall cavities. It settles in the walls and will leave your walls half empty. Use *batt insulation* in the walls with vapor barriers.

Ventilating your attic sufficiently will save you money in energy cost as well as roofing material cost. Extreme heat from your attic, if not ventilated properly, may shorten the life of your exterior roofing shingles, as well as, the roof underlayment. *Vapor barriers* placed correctly in ceilings, walls, and under your slab will lock out moisture.

Again, stay with the tried and true vs. the latest and greatest when selecting finish materials, but when selecting appliances and equipment, I sometimes <u>think</u> just the opposite. Always stick with reputable manufacturers, preferably American-made equipment, because it's less expensive to repair and replace.

First, research the PROS and CONS. There are tradeoffs with every selection. For example, many local power companies, to reduce high energy consumption, have promoted the heat pump.

The heat pump is energy efficient in a wide range of weather conditions, but in my opinion, in the extreme temperature ranges of cold (colder climates) or hot (perhaps in the South along the coasts), it lacks the comfort of conventional systems.

In colder climates, manufacturers suggest adding heat strips inside the ductwork of heat pump systems to attain desired comfort levels. Heat strips are similar to burners on an electric stove. They are perhaps the least efficient energy-consuming device ever invented by man. So where's the savings?

On Southern summer days in extreme temperatures, I've found that you don't experience the cooling a conventional system affords.

I'm not a heat pump basher; I just want you to use common sense and investigate. Know what you are purchasing.

👍 At the completion of construction, you should receive ALL the warranties. Locate and read the ratings on panels attached to the equipment to verify that you have received what you paid for.

👍 Also, good planning and the right selection of materials and equipment could eliminate taxes that will otherwise be needlessly incurred over the life of your home. When was the last time you examined your utility bill?

I've seen taxes in excess of 15% charged on the average monthly utility bill, another good reason to insist on adequate insulation and ventilation systems when planning your home. Keep those utility costs down and eliminate excess taxes that will never abate.

Chapter 6

HERE'S THE GAME PLAN!

(Plans, Specifications And Contracts)

...the designer, the plan designer was not local, and he wasn't familiar with the local codes, nor were we. We brought this up to the builder early on, and he looked at the plans and said he would take care of it. We were told not to worry. We trusted him to do that.

Eventually, during construction, a local official brought to our attention that according to F.E.M.A. maps, we were in a flood zone marked A9. Therefore, we were required to have within one foot of the grade standard approved surge vents. I think that's the term they used. We weren't informed about this until after the framing of the upper floors was already in place. It wasn't the builder who informed us of this. It was a code enforcement officer.

It's a F.E.M.A. requirement that the builder apparently didn't know anything about. The lower chain walls were in place, up underneath the house, and the floor decking was on. The chain walls are made of concrete, up to about six feet. So, rather than go back and do it right by cutting into the concrete and installing the vents at the proper levels, the builder just put them up about six feet above the ground, which didn't achieve the same purpose. He left a large open space in the walls for entry and access. We also had to install another door in that wall along with a standard six-foot door. They put two of the surge vents in that door, and, according to the builder's superintendent, that's all we needed. The F.E.M.A. inspector let him get by with what he had installed. The vents are located five feet higher than they are supposed to be. They are basically useless.

The misplaced vents ended up affecting our insurance. His error effectively doubled our flood insurance premiums for the life of our home.

We brought this up to the builder, but it went nowhere. We're still paying double and will be paying double for as long as we own this house.

RUNNING THE TRAPS OR WALKING THE TIGHT ROPE? THAT IS THE QUESTION!

Listed above in "Plans And Specifications," we defined terms and discussed where to obtain many of the documents that will ultimately play a role in the design of your home. Further, your construction documents must *conform to local building codes, restrictions, and zoning ordinances.*

Irregardless of whether you choose to engage an architect or designer or decide instead to purchase off-the-shelf plans, make sure that your contract for services concerning the plans and specifications states that "all plans and specifications shall comply with any and all local governing bodies including but not limited to building codes, zoning ordinances, restrictions, and subdivision covenants."

The professional that prepares your plans will consult your subdivision plat, a recent survey of your lot, a title report prepared by a title company, if available, the homeowner's association covenants, the local building official, the local zoning department, fire department, county drainage authority, marine resources, local governing bodies administering wetlands protection, and any and all other governmental, quasi-governmental and regulatory authorities having jurisdiction over design and construction in your area.

A survey of your lot coupled with your title report may indicate front, rear and side yard set backs, easements, and encroachments on the lot. *Homeowner associations (HOA)* make available subdivision covenants and deed restrictions. Design committees elected by the HOA may ultimately have authority to approve or disapprove your plans. Such things as appropriate use of exterior materials and colors may be outlined in the covenants. The minimum allowable buildable square footage may be defined. Time to complete construction may be defined. Roof slopes and maximum height restrictions will be found in the subdivision covenants, zoning ordinances and building codes. Placement of garbage-disposal devices, mailbox location, and off and on street parking may be designated in the covenants.

The United States Postal Service will also have a voice in locating your mail drop and designating your address, along with emergency services 911.

Caution: If there is no active homeowner's association, check the local courthouse records and have an abstractor from a title company verify that there are no subdivision covenants or deed restrictions filed for your subdivision. The local *building official* has the authority to grant building or construction permits that meet minimum building code requirements adopted by the city or county. There are many codes including: C.A.B.O., S.B.C.C.I., U.B.C., International Codes, etc. Check with your local building official to determine which building code applies in your particular location. Building codes and zoning ordinances also outline minimum yard setbacks, roof pitches, allowable building materials, construction types, etc. Your local *zoning department* has adopted ordinances outlining height restrictions, the type of land use or occupancy allowed on your lot. Geological surveys may be required. Historical committees, usually considered a subset of the local zoning department, oversee designated historical areas. Check with your local zoning department to determine if your lot is located within one of these *special use areas.*

Marine resources, county and city drainage authorities and the Corp of Engineers, especially in coastal areas, govern issues related to wetlands, flooding, and drainage. Check with your local building official and area zoning agency to determine who enforces drainage, wetlands, and flood surge requirements.

F.E.M.A. (Federal Emergency Management Authority) maps indicate flood prone areas. A local appraiser or your county engineer may have copies. Surveyors will indicate on your survey flood zones that may encroach your lot. If your lot or any portion of your lot is in a flood prone area, flood insurance will probably be required by your lender.

Topo survey, topo, or topo map are acronyms for topographical survey. A **topo survey** indicates the elevation variations that occur on the surface of your lot. This survey is used to determine the cut and fill required of your lot to locate your final slab height. Check with the local county engineer for regional topo surveys. Large scale topo surveys commissioned by local governing authorities are becoming more prevalent due to wetland and flooding issues. They might save you money if you can locate one.

Surveys come in several varieties; one, already described above, is the topo survey. You can commission your surveyor to shoot a topo of your lot and place it on the boundary survey for your convenience. You'll need a **boundary survey** outlining the perimeter of your lot in order for your designer to develop the site plan for your house. Usually, a topo is an inexpensive addition when prepared simultaneously with your boundary survey. When I order a topo, I usually specify that elevations be shot on a minimum twenty-foot grid. That usually supplies sufficient information for site planning. Of course, let your design professional order the surveys because there are many other items he may require of the surveyor, including but not limited to locating the trees on the site, their trunk sizes and types, locating the top of the street curb, the crown of the street, any storm drains, manholes, utility flow lines, fire hydrants, etc.

Another type survey is a **slab survey**. Your banker may be the first to request this survey. Once the slab is poured and set, the lender will ask that a slab survey be completed. The slab survey documents the location of the slab on the site to make certain it isn't encroaching into yard setbacks. Often, before it is poured, the builder will ask the surveyor to stakeout the slab to ensure it is formed in the proper location on the site, thus assuring against error.

Your surveyor can stake the corners of your lot, a **staked survey**. Use this to layout perimeter improvements such as fences and locate lot encroachments.

Your architect, designer, or builder will know who to contact for discovery and surveying services. *Information used in making design decisions should always be in written format. Mail summarized phone and conference meetings for review so that any misunderstandings can be corrected.*

When reviewing codes, restrictions, and ordinances, understand that the most restrictive will always govern.

👍 Precautions should be taken when purchasing **off-the-shelf-plans**! Have a professional examine them to *determine if they conform to local building codes.* Just as a local architect or designer makes a code review to ensure that their plans and specifications conform to local building codes, you must do the same for off-the-shelf-plans. A professional can help you adapt them to the local codes and ordinances. Usually, the owner of such plans encourages you to check with local authorities to ensure that plans comply with local codes. Authorization to make changes necessary to bring them up to code is granted without hesitation or additional fees. It's less costly to update plans than to correct construction not in compliance with codes.

HOLDING IT ALL TOGETHER OR ARE ALL SLABS CREATED EQUAL?

Commission a local engineer to design your slab. In the "trade," slabs designed by engineers are often referred to as **engineered slabs**. A professional engineer (P.E.) licensed by the state in which you intend to build will examine the loads, the weight of the house, and the bearing capacity of the soil to determine the size and location of any necessary grade beams, piers, or other devices warranted due to soil composition. His service is relatively inexpensive. An average slab design should cost you well under $1,000.

Also, have your lot's soil tested. The test will determine the soil type and also indicate its bearing capacity; how much weight it will support. These tests will eliminate guesswork concerning the type of foundation required. It's well worth it. Do it! Ask your engineer or architect which testing company he or she recommends.

Some *master builders* familiar with soils in your area will visually examine the soil to determine a slab type. Some might conclude that on the average they have been lucky. But to safeguard against a cracked slab, have your soil tested and have a local engineer or architect design your slab.

A BIT ABOUT WOODEN THINGS—NOT NICKELS!

Trusses eventually sag, and when selecting floor coverings, you must take this into consideration. Brittle flooring finishes such as tile may eventually crack due to truss sag. *Dimensional lumber* normally holds truer to form. Dimensioned lumber is just what it implies. Lumber milled to industry standard dimensions such as 2 X 4's (1 ¾" X 3 ½") or 2 X 12's (1 ¾" x 11 ½"), etc. A knowledgeable builder will disclose upfront material nuances to you.

Normally, use 2 X 4's for studs in the walls up to two stories in height. With additional stories, lower level studs should be increased to 2 X 6's. Usually, spacing the studs 16" on centers is recommended depending on the loads in the walls. Wider spacing is not recommended. Extreme wind forces or extra loads may call for closer spacing and larger studs. Hurricane and earthquake prone areas require special construction techniques. Your local codes will establish minimum standards.

For more information, I recommend the *Pocket Size Carpenter's Helper* as an excellent reference. It is concise, superbly organized, and something you can carry in your pocket when visiting the job site.

Chapter 7

PUTTING YOUR PLAN INTO ACTION!

(Getting It Built)

...as a builder, he has a great ability to pass the buck and blame others for his inadequacies. He also has the ability to talk a good game, but lacks the knowledge, ability and desire to produce what he promises. We went into this agreement expecting an honest, professional business relationship. What we got was never knowing from one day to the next if we had been told the truth, the partial truth, or the final truth. Instead of a pleasurable experience in building our hopefully last home, we ended up gritting our teeth, praying for the end to come, which is where we are today.

AUDITIONS ABOUND: HOW TO FIND A BUILDER!

Finding someone to build your home won't be a problem. *Finding a qualified professional is going to take some effort.*

Your first thought may be to build it yourself, which I wouldn't recommend. The following story about a simple pool-house may enlighten you enough to explore the alternatives.

I designed a simple rectangular pool-house for a very savvy couple. They were outgoing, knowledgeable, and astute. They assured me they could handle the construction.

"No problem!" they exclaimed.

They'd done it all before. After a few visits, I got to know them better and felt comfortable that they could, indeed, oversee their addition, a simple structure. After all, they were, in fact, supervising the remodeling of their own massive historic home, and quite well, I might add.

Months later I had occasion to visit with them. Their first expressions alerted me to the fact that construction hadn't gone according to plan.

The contractor they trusted had poured their slab out of square. It wasn't discovered until he had been paid and was off the job. In fact, the next subcontractor, the framer, quickly discovered the problem when attempting to square the first wall corner.

It came down to this. Unless they tore out the slab and started over, the framer would have no alternative, but to tackle the challenge of framing an oblique pool-house, a skewed parallelogram. Every trade thereafter, would have to adjust to the same problem. The owners were distraught, but there wasn't much they could do.

Someone once put it best, "Whatever can go wrong, will go wrong." The lesson learned is that you can't leave subcontractors unsupervised. Check their work plans before they commence, their work during construction and their work upon completion, and <u>before</u> they are paid. This example also illustrates the critical nature of placing the slab correctly. In <u>so</u> many ways, a house is only as good as its foundation. Hopefully, recalling this little snafu will help anchor you in reality. Magnify it a hundredfold, and you may begin to visualize the complications of a skewed house slab.

With that aside, it's time to explore other alternatives to building your dream home. For purposes of our discussion, **builder** is defined as being one of several disciplines capable of managing your home construction.

The obvious choices will be one of the following three: a *construction manager (CM)*, a *home builder* or a c*ontractor (general contractor or prime contractor)*. The home builder and the contractor are considered prime or general contractors in the sense that they contract and coordinate the subs taking full responsibility for their work.

The construction manager does not provide this division. Simply put, the prime stands between you and the subcontractors and vendors, possibly limiting your liability. The construction manager does not, unless so authorized. Further considerations will be explored below.

The first real choice, in my opinion, is the **construction manager**: an architect, engineer, home builder, or other qualified experienced construction professional who is educated, as well as, experienced and trained in the proper ways and procedures of construction.

The construction manager's (CM) scope of work may include managing contracts, managing subcontractors, scheduling construction, developing a construction budget, establishing a completion date, and adhering to it. His fees will be based on either a percentage of the total cost, a flat fee, or an hourly fee. 👍 As with any other construction contract, make sure that he is tied to your completion date and budget realizing that his performance will be dictated by contracts he negotiates with subcontractors. 👍 Review all contracts before they are signed to determine whether their timeframe and costs mirror your timeframe and budget. In fact, you might want to make all contracts contingent on your final approval and signature.

👎 Usually, the downside of using a construction manager is that there will be no *prime,* general contractor or builder, who assumes the risk. Using a prime contractor is the traditional method of construction.

Normally, you would sign a prime contract, a master contract, with him and then all subcontractors and vendors would in turn be contracted by and to him. Subcontractor or vendor disputes would typically be resolved by the prime. Using a CM eliminates the prime and each subcontractor and vendor may, under the law, become a prime if contracted directly with you or your CM. In certain states, under this arrangement subcontractors may be entitled to lien your property directly if there is a dispute over payment or work product unless you *contract around it.* How do you accomplish this?

Your CM may require in writing that subcontractors relinquish all rights to place **liens** or **lis pendens** (a recorded document as notice of ongoing legal action) on your property. It's a good idea to use this tactic regardless of which method you eventually choose for construction.

One last point while on the subject of liens, each prime when paid must certify in writing that all vendors and laborers have been paid "in full" for material supplied and work completed.

Check with your prospective construction manager or local title or real estate attorney to determine how your state laws might affect you in these matters. When hiring a CM, make sure you engage someone who is specifically working in home construction in your area. He will have firsthand knowledge of subcontractors and vendors for better pricing.

Understand that a standard CM agreement will NOT require the CM to finance your construction through the first draw as a prime (home builder or contractor) normally would. Remember, with a prime contractor, payment for work is made at the completion of each phase, only after it is completed. The CM's main purpose is to manage construction through to completion. He is a paid employee, agent or consultant. Therefore, any immediate upfront cash requirements will be an owner responsibility. That's you!

When employing a construction manager, you'll be required to set up credit accounts with local vendors for materials. For the best prices, you will also be required to meet weekly payroll for subcontractor labor.

The benefit, everything is at your fingertips. You're in control. The downside, everything is at your fingertips. You're in control.

The next traditional method of construction would be to use a **home builder**. They are, by far, the most prolific builders of homes. Start your research by calling your local home builder association to receive a membership directory. From it you will learn little about the background of their members, their credentials, their qualifications, or whether or not they have any satisfied clients, but it's a start.

Further your research by locating finished houses that you admire and inquire about the builder. Recent homeowners are notoriously happy to replay their building experiences. A good time to ask if they would indeed recommend their builder will present itself. Another good source for references is local suppliers.

Home builders come in all shapes and sizes. Before you contract, complete your due diligence and inquire at length about those you assume might be good prospects. Also, you may find better pricing with a home builder because they're always looking for the next project to keep their subs busy in the field.

The next alternative would be a **contractor**. We'll define a contractor as someone who engages in commercial as well as residential construction. A contractor more prone to commercial construction may have developed higher standards and use better subcontractors than a typical home builder. Along the way he has had to satisfy not only owners but architects who have helped educate him, especially about quality. His finished product may be of a higher quality, but his ultimate price may also be higher. Remember this, commercial construction is usually much cleaner than residential construction, and once introduced to commercial construction, a contractor may prefer to only engage in that type construction. Knowing this, don't let your project fill a lull in his commercial construction schedule. Make sure he is currently involved in residential construction.

You may consider using a builder coupled with a construction manager. This arrangement is common in large commercial construction, but of course, you must weigh the expense. If you are willing to stay the course, it may not be cost effective.

When searching for a builder, make sure you compare apples and apples—like homes with like homes.

Basically, there are three types of construction projects to be found in the marketplace. Talk to the owner or builder to verify which type you are inspecting.

The three types are *custom home construction*, *speculative home construction (specs),* and *turn-key or semi-custom construction.*

For our purposes, **custom home construction** is defined as a home designed specifically for an owner and built on his pre-owned lot. A **spec** home is built on a builder's lot as inventory while a **turn-key** project is built for a specific customer who may or may not own the lot. The turn-key usually takes on the form of a typical builder's floor plan retrofitted to suit the owner's needs.

Spec construction commences even before a buyer is in sight. Turn-key and custom houses commence after a builder has contracted to build a particular home for a particular individual to his specifications. To the unknowledgeable, the differences between owner and builder owned lots may seem insignificant, but in reality, it can make all the difference in the ultimate cost or leverage that may be required in adversarial situations.

Never compare custom home construction on your lot to spec (speculative) or turn-key projects built on a builder's lot. Spec and turn-key projects are handled much differently than custom jobs, especially when it comes to completion time. With little or no scrutinization of construction, the builder completes these spec jobs much faster. If you are carrying the construction financing and haven't managed to tie down the completion date or total cost, the builder will have a tendency to use your project as a filler between other more pressing jobs. You will, therefore, experience unnecessary delays. Time is money!

When exploring the PROS and CONS of each type, our illustrations will be kept clean and simple, but there are hybrids of each. Turn-key projects emerging from specs during construction would be one example.

The spec is the easiest to assimilate, simply put, you pay for what you get.

The builder owns the lot and the improvements and offers the home at a price. If you like what you see and feel the price reflects value, it's yours for the taking. It's that simple. The CON may be the standards to which the house was constructed. Another CON may be the lot, its location, quality of life it affords or the subdivision itself. Watch out for flooding back yards!

The spec is the simplest from which to choose. You want it, or you don't! It's easy to walk away from. Some builders even drag their feet in finishing a spec to allow the purchaser a few selections, normally a few finish materials and colors.

The next type of construction, the turn-key, is a little more complicated. The purchaser may participate in selecting the floor plan as well as modifications thereto, the lot, materials, and colors, and he should have more negotiating room when it comes to price. The purchaser may have more control concerning the outcome of the finished product, where it will be located, what goes into construction, and how much it will ultimately cost.

Two of the biggest concerns with this type

construction are construction financing and lot ownership. Before we consider both, there's one point I need to make.

The turn-key project presents the builder with a golden opportunity to CON. He may insist that you put up a deposit. His argument: he is building a semi-custom house to your specifications. He will argue that your selections are unique, and if you eventually back out of the contract before he finishes, he may never sell the home.

Never put down a deposit, unless it is a token amount! The PROS and CONS of deposits will be discussed below. But for now, proceed with caution. Remember this, never pay for work that is not complete.

When considering a turn-key home, you may be asked to secure the construction loan. I would advise against that, unless you own the lot. Realize this! Eliminating any financing obligations will enable you to break clean if you decide not to follow through and purchase the finished product.

Further, to accomplish a clean exit, a **termination clause** must be part of your construction contract.

Reasons for breaking a contract are numerous, but wording to accomplish termination may be as simple as, "The buyer maintains sole discretion, as to whether, he will accept the final product. At any time upon written notification to terminate this contract, the buyer will have no further obligation, and this contract shall immediately become null and void." Now, let's explore a situation that might cause you to terminate a contract.

A turn-key home was built using I.C.F. units, concrete filled polystyrene blocks. The builder obviously wasn't paying attention, and the windows were inadvertently, incorrectly located. After being informed of the error, but without consulting the buyer, the builder jack-hammered the walls and with a chainsaw and sledge hammer finished opening up the rest of the concrete reinforcement to relocate the windows to their proper position. Once notified, the buyer surmised that the structural integrity of the walls went the way of the sledge hammer and he became irate with the builder's actions and demanded out of the contract. Reasons to void or disengage from a contract are too numerous to elaborate here, but the above scenario should lay the groundwork for further contemplation.

Participating in construction financing adds complications, and can make it impractical to walk away from the project even though attorneys may caution you to mitigate your losses. That goes for the builder also. When sabers begin to rattle, your builder may elect to pull off the job, bringing work to a virtual stand still. He may be cutting his losses as advised by his attorney. Yet, construction interest payments still accrue. Payments made by you!

Hence, interest payments may work in the builder's favor if things go south. Sooner, rather than later, you may be begging your builder to again commence work; perhaps, automatically positioning yourself for more abuse. That's hard to digest, but it happens more often than you might imagine.

Construction interest can be significantly higher than a normal mortgage note rate; large interest payments ensure that time to complete construction will become a major consideration. The completion deadline must be defined in your construction contract. Stay away from construction financing, unless you know your builder, and have taken the time to prepare a concise construction contract; thereby, limiting your exposure to interest payments.

The third type of home construction is custom home construction. You own the lot, and you simply hire a builder to construct the house according to your plans and specifications.

Know your builder! Do your homework by investigating your builder! It will pay off. To your chagrin, because you own the lot, arguably this may be considered the riskiest type of project.

You will be faced with a choice; secure construction financing yourself or leave it up to your builder. If you want the tail wagging the dog, let him secure construction financing in this situation. You will subordinate your lot ownership to his construction loan lender. I would <u>not</u> consider that to be an astute business decision. It puts him in control of your destiny.

Complete your due diligence by researching your builder's reputation and don't ignore what you find. "Where there's smoke there can be fire." Your contract for construction must be in order, with every "i" dotted and every "t" crossed. The right builder and the right contract will help ensure your ultimate success. But if you've been unlucky and chosen "Slick Eddie", problems may start to occur from the onset.

Lot ownership and construction financing may begin to loom as large negative forces that checkmate you into an undesirable disadvantage. To remedy, the only way out may be to discharge your builder.

It becomes exceedingly difficult to walk away from construction on your own lot. Legal ramifications surrounding your construction financing will complicate issues and make it almost impossible to walk away, although, I've seen it happen. Here's an example.

A builder contracted with a couple to build a house, a typical wood-framed home. The couple owned two lots and planned the home to straddle both lots. They secured construction financing. During the negotiations, they elected to use I.C.F. units, concrete walls the builder was pushing, instead of wood framing. Due to cash flow problems or pure negligence, the builder got ahead of himself and started the job without reworking the plans. Neglecting to have the plans redrawn, to reflect the increases in the wall thickness created by the I.C.F. units, would prove disastrous. When the house was finished, some of the room sizes were reduced due to the additional thickness of the walls, and the resulting spaces could not easily accommodate furniture.

The builder insisted it was the owner's fault for choosing the I.C.F. construction. According to the owners, the builder even forged their names on a set of plans trying to convince them that they had agreed to what he had constructed. Attorneys entered the picture. The builder eventually consented to buy back the house and the two lots.

That was an unusual ending to a problem that to often plagues the business, builders with no accountability or with "disposable integrity". The couple recovered most of their money, but lost their prized lot location, time spent rectifying the situation, not to mention the bad experience which still haunts them to this day, along with attorney costs.

Above all get references. *Asking for a client list can be a real showstopper!* If a builder has a problem furnishing a list, don't hesitate to discard him immediately.

Some builders would make you think they are beyond reproach. Don't believe it. CHECK THEM OUT, no matter how many homes you see them building! The time is NOW before you contract with them. Take this one step at a time. Inquire as to where they have built and talk to past customers. Don't become a victim of naiveté.

Just because he gives you a few customer names, doesn't mean you shouldn't check them out. I'm not just joking. Check them out! Really!

Honest people assume that people with whom they associate will also be honest. *I've more often than not, found this to be untrue.* What you find in your due diligence period may shock you.

The success of your research will determine whether you'll enjoy an exciting and pleasurable experience or a grueling, nightmarish one where you can't wait for it all to end!

Once you've narrowed your field of builders, check out their **superintendent** (super), the person who will directly orchestrate your home construction. He's your contact. He should be a body of knowledge. Quiz him or her!

Questions might include: Where did he receive his education and in what? How many superintendents does the builder employee? How many houses does the builder complete in a year? How many jobs does your super presently handle (supervise)?

It's like determining the ratio of teachers to students. The more teachers, the more attention each student receives. Also, inquire about the proximity of his present construction projects to your job location. After gathering this information, ask yourself, do the logistics make sense? Will he be able to handle my job? Logistics are a major consideration. Once the builder has contracted with you, you may never see him again. It's then up to his employees to get your home built.

Home construction demands much coordination and supervision. It doesn't happen on its own, and it won't happen from afar. Use your common sense in making your final determination. Make sure you gain some knowledge about your superintendent's education, training and experience. If you like him or her, designate him or her in your contract as job site superintendent.

Have your own representative talk *shop* with him to ascertain how much he really knows about construction.

You might be surprised at his lack of knowledge.

Scratch the builder off your short list if he shows resistance to your questions.

Remember the builder isn't doing you a favor by building your home. It's how he makes a living. It also pays to remember, that some small builders look at custom or semi-custom jobs as a bird in the hand, an opportunity ripe for plucking. Proceed with caution!

Chapter 8

WRAPPING IT ALL UP IN A NEAT LITTLE PACKAGE!

(Budgeting)

...not having a budget made me as nervous as a NUN on a honeymoon...

CHOOSING OR LEAVING IT UP TO THE OTHER GUY: WHO CONTROLS THE BUDGET?

...we *were intent on letting this builder who just happened to be the president of the local home builder association erect our home. A consultant had estimated the cost at well over $600,000. Without running the necessary traps or checking out any of his past clients, we contracted with him due primarily to his standing in the home builder association. Ultimately, we were surprised and disappointed!*

We ended up being one of those couples who gritted our teeth, hoping for the day it would all end. Later, a third party who prepared cost estimates for the builder confided in us. He discovered that the builder didn't even have the knowledge to develop a true cost estimate. In fact, the builder had been overheard telling his wife that he would have to get out of the business if he ever had to develop a firm cost estimate.

Never execute a construction contract until you have agreed on a **stipulated (firm) sum**, a total cost.

As explored above, in the early design stages, your budget will begin as a simple cost-per-square-foot serving as a ballpark estimate at best. If the builder quotes it, remember it, write it down and date it and make sure it is based on the construction style you are considering. It may be a *low-ball* cost estimate, one low enough to keep your attention and keep him in the game. Refer to it throughout the design stages and price negotiations. It may help to keep the builder honest. At some point, a line item cost estimate should be prepared by your builder especially if you intend to work under a cost-plus contract.

As we discussed previously, cost-per-square-foot, a type of unit cost, is dangerous if used without understanding what it represents. Besides, there are many factors that must ultimately be considered when developing a true budget, for instance, linear feet of wall space.

The object is to reveal the actual costs required to build the heated area under roof. At first, all other costs within the cost-per-square-foot price are irrelevant. Determine your actual house costs before concerning yourself with lot amenities.

Remember, if land is included in the cost-per-square-foot quoted, eliminate it. Subtract any other costs including garage costs, auxiliary outbuilding costs, amenity costs, extra costs, which may include a pool, driveway, expansive parking areas, landscaping, etc.

Once you have eliminated all extraneous cost from the cost-per-square-foot, you have effectively revealed **cost-per-square-foot** to build *the heated and cooled area under roof.* That's the closest you'll come to costs until your builder commits, or you commission a complete cost takeoff or prepare it yourself. And, remember, all cost takeoffs are a combination of science and art. *Further, a cost takeoff never really represents true cost until someone inks on the dotted line to build your home for that sum.* Cost-per-square-foot to build heated and cooled space represents a very rough cost multiplier to be used cautiously in ball-parking early design conceptions, and that's its limitation.

Design considerations affect final costs. These include the linear feet of exterior wall space, the cubic footage of interior space, the total amount of square footage, types of finish materials, etc.

Establish early cost estimates by visiting past clients of your prospective builder and asking them about the cost and size of their homes. Go through the same process outlined above to reveal the cost-per-square-foot of heated and cooled space.

Keep the following general rule of thumb in mind. The greater the square footage, the larger the home, the less it will cost-per-square-foot to build even though the overall cost will be greater.

Why? It's simple, the greatest concentration of costs remains in the bathrooms and kitchen. If you maintain the same number of baths and same appliance mix, cabinets, and counter space in your kitchen, but increase your overall room sizes, the overall average square footage cost will drop. Averaging the largest concentration of cost with more total square footage of house lowers the overall cost-per-square-foot. But remember, compare like homes.

A sure method of determining real cost is to ask the experts; the architect, the designer, the builder, cost consultants, or your personal representative. Give them all a shot at it and compare notes.

Design professionals must give much consideration to cost when developing your construction plans and specifications. Your designer can't dictate to your builder what your home will cost, but his knowledge of costs can present a strong argument in your favor; therefore, helping to keep your builder within the lines. Pre-selecting your builder will allow him to assist and complement your designer assuring that you stay on target and within budget. Remember, earlier you established the maximum you will spend, your budget, which included the maximum loan amount and cash you intend to use. Including your builder early in the design process saves a lot of time and money. It should eliminate the need to rework plans for budget purposes. At each stage of plan development, have your builder update his cost estimate.

When plans and specifications are near finalization, the builder should prepare a *line item list of costs*. He might present you with a schedule of values; a list prepared for the lender. Remember, this is a checklist the lender will use to advance **draws**, installment payments, to the builder during the construction process. The schedule is a compilation of all costs (values) broken down into job subcategories such a slab, roofing, etc. Have your builder break down the subcategories and give you a vendor list itemizing the quantity of material estimated (quantity take-off), labor costs, and costs from each vendor consulted.

The vendor's list should include the costs of all appliances, cabinetwork (casework), fixtures, lumber, sheetrock, etc. You'll want to verify that items on the list reflect those identified on the plans and in the specifications. To complete your check of the cost, which will protect you in the coming months from *price gouging*, have your builder provide you with a schedule of costs. This will be in the form of unit prices and will include cost-per-square-foot to install such items as bathroom floor tile (wainscot and floor), cost-per-square-foot to paint, to sheetrock, to frame the house, to set cabinets, to trim (hang doors, hardware, wood trim), etc. A schedule of labor costs and or unit costs will serve as a guideline ensuring fairness in the months to come. You will reference it in your contract and use it to add or subtract changes, but you must reference it in your contract.

Remember, each category of construction has related labor costs and material cost. Some have multiple layers. Take, for instance, a roof. It may include decking, black paper, roof shingles, nails, metal flashing (gutters, drip trim, valleys), and perhaps storm guard. The work just described usually calls for two different trades.

Again, costs should be no secret at this stage. If the builder is knowledgeable and reputable, he should have no problem supplying the necessary costs of the items specified. If it's a *lump-sum or stipulated-sum* contract, a contract stating the total cost to build the house, he'll want to make sure he has included everything in his price. Now is an excellent time to review those costs.

I recommend *third party cost estimates.* Disinterested parties, consultants, can provide you with a good cost breakdown for a reasonable fee. Local firms provide these services. Find them in your phone book or ask your designer to recommend someone.

Another method of determining cost is to calculate them yourself. *Go directly to the source.* Your local lumber company will prepare a complete cost estimate of the materials they can supply. Vendors that stock items specified in your plans and specs will have costs and cut sheets for your reference. They will also supply you with the quantity of materials required to complete the job. This becomes an important accounting tool as the job progresses. Note, your builder should be able to "best" your prices because of his work volume and discounts he may receive for promptly paying his bills.

Securing labor cost will be a little more challenging. Again, on the whole, your builder's costs should be lower than what you will find in the marketplace because his volume of work will afford him special prices from the subs. He will also take advantage of other inherent opportunities of the business. Picking up subs who are in between jobs is a good example.

Unless you intend to use large commercial subcontractors who maintain offices, you will have to find residential subcontractors (subs) in the subdivisions where they work. I'm the first to admit, some of these guys look a little scary after a hot summer day of work, but most are just hard-working people making a living. Find laborers (subs) skilled in the jobs necessary to complete the work on your house. Always make sure you are communicating with the boss. Don't panic! There aren't that many trades involved in building a house. Show them your plans, but don't leave the plans with them. Subs are infamous for losing plans.

Subs are usually very accommodating. Most subcontractors base their prices on a unit price, either cost-per-square-foot, or *square* which is a hundred square feet. The cost-per-square-foot may be floor area, wall area, board (e.g. sheetrock boards) area, etc. Asking questions will ensure that you understand how they arrived at their quote.

Ask them for a material take-off, an estimated quantity of materials. Check it against your vendor's quantity take-off. It's usually not a good idea to let a sub supply the materials. That will cost you a handling fee.

This process can be challenging, especially without a good construction vocabulary. Don't let it get the best of you. Checking with more than one sub concerning the same job will allow you to cross check their work, learn some vocabulary while comparing prices and quantities. Subs will work with you to make sure you cover all costs. But you'll have to keep calling them. I assure you, everyone does, even the builders.

A few gaps, cost deficiencies, shouldn't be significant. Of course, your representative will be an invaluable resource in this process if you've elected to use one.

A few costs you probably won't think to include are trash removal, temporary power, potty, clean up, permits, and insurance.

THE JOY OF NOT PUTTING OFF TODAY WHAT YOU CAN DO TOMORROW: AM I TOO OLD FOR AN ALLOWANCE?

An **allowance** is a cost that may be pulled out of the hat to represent some item that hasn't yet been selected. To arrive at a final cost, the builder simply throws in a number, realistic or not. It could be a WAG, a wild ass guess. I recommend before contracting that you find a vendor and select ALL items not specified.

Never allow a line item cost to appear as an allowance. Select every item. Firm up the cost and leave nothing to chance. There'll be enough decisions to make once construction is under way.

COMMITTING OR MERELY HINTING AT SO: WHAT IS A COST ESTIMATE?

A cost estimate is simply that, an estimated cost representing an opinion of what it will cost to build your house. A cost estimate remains an estimate until someone commits in writing to use that cost figure as a total sum cost, lump-sum or stipulated-sum to erect your home. They all represent the same thing: a firm price to build.

Chapter 9

MEANS OF GETTING TO THE SAME END!

(Standard Procedures)

Negotiating the contract turned out to be quite a hassle. The builder finally convinced us to go with a "cost plus" contract. We could "save money" with it. We were to pay for the cost of work performed, plus a predetermined percentage on top of the cost to cover his profit and overhead. He explained that we would only have to pay for the actual cost of the work; therefore, he wouldn't have to inflate his estimate to cover any unforeseen costs. It sounded like a fair deal. It seemed logical. If there were any cost savings, we would benefit.

The builder had just been paid for pouring the slab when the bank called asking me for the "slab survey." I wasn't sure what a slab survey was before I called my surveyor and ordered it. I soon learned that it was used to identify the location of the concrete slab on the lot. The bank uses it to verify slab height and also to make sure that the slab has been placed in the right position, not encroaching easements or yard setbacks.

To our dismay, our slab had been placed incorrectly, about nine inches into a "side yard setback." At the time I didn't fully understand the significance, but the banker sure did. He insisted that the builder raze the concrete slab and re-pour it, placing it correctly on our lot. I couldn't help but think what a fool this builder was, such a waste of time and money.

When the builder sent me a invoice for razing the first slab and pouring a new one, I went berserk. There was a note attached to the invoice reminding me that my contract stated that I was to pay for ANY and ALL cost (apparently even errors) plus a percentage of profit and overhead. The guy even had the balls to bill me for profit and overhead on his error!

CINCHING YOUR DEAL: SOME CONTRACTS ARE BETTER THAN OTHERS.

Understand one thing. If nothing else, leave with this. There is only one legitimate type of contract in construction, and that is a *written contract.* That can't be overemphasized! A **written contract** includes the original construction contract referencing the plans, specifications (construction documents) and any *change orders* and *field orders* that may occur during construction.

Everything must be in writing. Always reference your plans and specifications in your construction contract including the date they were issued, the number of pages, and the issuer.

Generally, construction contracts come in two formats, *lump-sum* and *cost-plus.* The CON is the cost-plus contract. The lump-sum contract is sometimes referred to as a stipulated-sum. It means firm price. The cost-plus contract may be divided into two categories, the cost-plus with a firm price or ceiling and the cost-plus.

Lump-sum contracts are the only style contract I recommend, especially if you haven't logged serious time in construction. The other types of contracts, which we will discuss below and have already mentioned above, are dangerous in the wrong hands. Take it from the casinos that have become so popular throughout this country; the odds are stacked against you.

The **lump-sum** contract simply states that the builder will erect your house according to your plans, specifications and contract for a specific sum, firm price.

There are no ifs, ands, or buts to this contract if

you've followed my insights. Change orders should be the only instrument that can change the price of a lump-sum contract. The change order should be authorized by you in writing before any change occurs. This will allow you time to review and accept any cost change or contemplated deadline extension request. Make sure you explore this completely in your construction contract.

If the builder won't sign a lump-sum contract reflecting a realistic firm price, then there is a hidden agenda. Either he doesn't know what he is doing, or he does. **You lose either way!**

Inform the builder at the onset during the initial stages of plan development that you have decided the lump-sum contract is the only contract you will sign. Don't let him talk you into a cost-plus contract. Communicating with some of his past clients might help you anticipate what type of contract he customarily prefers. If he is NOT willing to sign a contract stating the firm price, once plans and specifications are completed, you've got a problem. I would consider dismissing him before wasting any more time.

The next style of construction contract, which I don't recommend, is the cost-plus. It evolved under the premise of saving the client money. Only a construction-savvy client should use it, and then only with caution, otherwise, it will cost you money. Unscrupulous builders use the mechanics of this contract to escalate costs. When used correctly by a reputable builder, it has its advantages, but for the general public, the possibility of abuse far outweighs its advantages. A safeguard when using a cost-plus contract is to insist on a **firm price or ceiling**, a maximum price the house will cost.

For purposes of this discussion, the cost-plus contract will be divided into two parts, the cost-plus and the cost-plus with a ceiling. The **cost-plus** portion of this contract simply states that the builder is entitled to a mark-up fee, a predetermined percentage of every penny spent on the construction of your home. You might think of it as a tax by the builder. You're taxed on everything you spend. This percentage must be negotiated beforehand.

Some builders even try to mark-up (include a profit) materials and labor furnished by the owner. This practice should be unethical. Make sure you address this in your original contract, or it will happen to you. Make no assumptions when using this type of contract. Read it carefully and thoroughly, understand it, and define everything.

👍When using a cost-plus contract, make certain it is tight and loophole free. If you want the freedom to supply material and labor for a particular aspect of the project and don't want to pay a mark-up fee on those particular costs, outline what material and labor you intend to supply by stating it in your original construction contract.

A realistic mark-up ranges from 10% to 20% depending on the market and what's involved. The mark-up should represent ALL overhead and profit. State that in your contract and define overhead.

Unscrupulous builders that prefer the cost-plus contract, use as ***bait*** an overall estimated cost that is low. After the ink is dry, there is no obligation on his part to honor that price. The main purpose of a low-ball estimate is to induce you into signing the contract. Once you're contracted, the sky's the limit on what construction will actually cost. I've witnessed the original cost estimate increase over 50%. Don't let it happen to you!

If you insist on using the cost-plus contract, there are some basic mechanics you must understand. First, define which items, materials, and labor your builder is allowed to mark-up. Next, define what the mark-up represents. Does it represent both profit and overhead or just profit or what? This is important. If it is not defined in writing in the contract, you're doomed to be gouged.

I've witnessed builders charging as much as 20% mark-up and then turn around and charge a huge additional overhead cost which should have been included in his mark-up. Overhead costs that may include several superintendents whom you never see, cleanup crews who never clean, off-site bookkeeping, etc. The builder can get really creative with costs if left to his own devices. Additional mark-up and superfluous fees are eminent unless otherwise defined in your contract.

Establish up front, in writing, what the mark-up represents. When using this style of contract, you must also define *who pays for construction errors and damages* to work completed or stored on-site. It's important to define in writing that the builder will pay for his errors (labor and material), defects and damages to materials, and especially the costs to fix if he doesn't build the house according to the construction documents. Allowing the builder to mark-up his time and materials to correct his mistakes caused by his own negligence or error is unacceptable. It's not prudent to reward him for his mistakes.

If it's a change you request, not an error, before you authorize it, demand a proposal from your builder. In his proposal he must itemize all costs for the requested change and any additional construction time extension necessitated by the change before you authorize him to commence. Make sure this series of events is defined in your original construction contract, that he document any change in writing and that it must be authorized by your signature on a change order before commencement of such work. You need time to review, before your final authorization. But whatever you do, don't proceed with the change without the proper paperwork. If you're not savvy about construction, you probably won't know or understand what other parts of the project your change might affect, thereby costing you money. Have him explain it. Have him put the total cost for the change into writing. Again, this is usually accomplished by formal documentation called a *change order.* Contact your local or regional American Institute of Architects (A.I.A.) or your banker for a typical change order form. Another item you should be aware of when considering changes in work is that the mere request for a proposal may increase overall construction time, if the builder slows construction in anticipation of queuing in your requested changes.

It's time to define the firm price or ceiling which must be an integral part of any cost-plus contract. It's simple; the ceiling is the MAXIMUM cost the builder is allowed to charge to complete your house. State that figure in your contract.

The ceiling protects you from price gouging. It isn't estimated; it must be a firm price. If his ceiling is a fair price it should be in line with the costs of a lump-sum contract. If it is in line, then there is a slim chance you may save some money. Understand what you are signing or seek professional advice.

Again, when electing this contract ***NEVER*** use an estimated ceiling, otherwise, it isn't a ceiling, and again the builder is under no obligation to complete construction for that amount. Remind your builder that the purpose of this contract according to him is to save you money. If there is any price difference, it should be downward. Therefore, insist on a price ceiling. Without a price ceiling, you are working without a net. A cost-plus contract must only be used with complete plans, specifications, and a thorough contract. The price ceiling must include every cost.

A further check is to compare the ceiling to the overall cost you have developed as outlined above. Do the costs agree? If not, why not? *Never engage in construction without a firm overall cost irregardless of the type contract you elect.*

The third and last style of construction contract, the pure **cost-plus** without a ceiling, should be *outlawed* in homebuilding. It's much too sophisticated for the unsuspecting general public. *I consider it a red flag indicating that your builder intends to fleece you.*

Never use a cost-plus contract without a firm price ceiling.

Your builder may insist on providing you prices based on two styles of contracts. If that's the case, you may notice that the lump-sum price is higher than the other two types. Why?

There could be several reasons. We will explore the two most common. Firstly, your builder might want to pad his costs to ensure his profits. But, if he has followed a realistic process in developing costs as outlined above, he shouldn't need to inflate his final number. Ordinarily, the only area in which his calculations should vary would be site work and then only under extreme circumstances. Unknown underground streams or old slabs buried below grade are good examples of extreme circumstances. In fairness, your contract should address unforeseen site circumstances. A change order may be necessary to cover those extra costs.

Secondly, if the builder hasn't done his homework and developed true costs, one way he can ensure his profits is to inflate his costs. If he can't produce costs, I would opt out of a relationship with him immediately.

Do your homework and make him do his! Keep him honest and stay in control.

INSIDER CURVE BALLS, LOW BALLS AND FOULS!

Forewarned is forearmed:

1.) When negotiating a contract, remember that your RIGHTS are up for grabs. In America, you are allowed to sign away ANY AND ALL of your rights and no one will care.

2.) Never allow yourself to be forced into signing a construction contract on the spot. Insist on taking several days to review any contract you are presented. You need time to include items you deem important.

3.) Impasses. To remedy, in your contract call for a competent third party to settle all disputes. I define a **competent person** as someone who is formally educated and has experience in the type of construction project in which you are involved. It's a good idea to designate that person in your initial construction contract, but make sure you are able to change your designee from time to time. Outline in your construction contract the method and timeframe to be used in implementing the process of bringing in this third person when a dispute arises.

He or she will settle all disputes, and his or her decision must be final. It's tough, efficient, and it works. Do your homework when selecting that person. Your choice must be versed in the means and methods of construction and not linked to your builder. Make sure he has his own practice and that his integrity will not be compromised by a promise of future business from your builder. He must understand your construction documents and the scope of work outlined in your contract.

Further, if in an insane moment, you decide not to go with that suggestion, I would consider scratching any reference to mediation and instead insert a **binding arbitration clause**. The results of a lengthy and costly mediation are not binding, which simply means, that the losing party may or may not honor the results. Binding arbitration ensures that the losing party will honor the decision.

When defining binding arbitration, you must specify how a mutually agreed upon arbitrator will be selected, the timeframe, and how the overhead costs of binding arbitration will be divvied up. No matter which avenue you choose, scratch mediation and substitute something to finalize issues at hand. Both mediation and binding arbitration are more expensive and more time consuming than the third party resolution mentioned above.

An impasse is another reason it's important to use a complete set of plans and specifications and a thorough contract along with considering other insights outlined in this book. Again, designate when, where, how, and in what timeframe arbitration will be implemented. In your original construction contract, address how, when, and what money will be escrowed, if any, when a dispute arises. Personally, I would nix both mediation and binding arbitration and elect the third party alternative as the swiftest method for resolving impasses.

4.) **SETTLING ISSUES IN COURT IS NOT AN OPTION!** Court is no place to settle construction issues. Our court system is adversarial in nature. In a nut shell, things get nasty. Issues central to your claims run amuck in a lawsuit. The outcome of alternatives enumerated in this guide, when the dust settles, could save you time and money. To quote a trite phrase, "staying out of court will insure that you'll be far ahead of anything in second place."

If there are winners, it would have to be the attorneys! Many of whom are uneducated, unknowledgeable, and inexperienced about construction, business, real estate, and financing issues.

It's frightening when your attorney doesn't even understand the meaning of equity, a concept basic to the understanding of real estate. Amazing you say? It's alarming to think that the more ineffective an attorney is, the more money he stands to make at your expense in defending what he puts on paper or doesn't put on paper. I should enumerate on this behavior in a future book.

Not all, but more often than not, I've dealt with attorneys drunk with ego, unprepared, predatory in their practices, exploiting people who found themselves in desperate situations, literally fanning the flames to assure additional fees instead of striving to resolve issues to an early, amiable and rational conclusion.

Some attorneys will oversimplify the issues at the point of engagement, but untangling construction issues is very time-consuming and expensive. It takes experts, which translates into lots of money. Most capable attorneys realize it will be hard to justify their time and fees, especially if the results are unfavorable.

You must find a capable attorney (firm) with experience and knowledge in the laws pertaining to the issues, and he must be willing to devote enough time to be your advocate. Like personalities, attorneys come in all types.

Reflecting on a short story once told to me years ago by my son who had heard it from an attorney himself, may help focus your awareness.

One day a young promising attorney discussing and scheming a little with his mentor about his plans after graduation said, "I was thinking. Should I settle in a small town with no other attorneys? I imagine I would do rather well."

Amused, his mentor, a wealthy attorney, smiled and replied, "You'll starve! Find a town with at least one other attorney. Then, you'll have the makings for a real financial fortune."

This brief story illuminates a serious concern about that profession, but will it galvanize your focus enough to consider alternatives to court when considering construction issues or any other business issues to be addressed in your contract?

Design and construction issues are complex and can be overwhelming. Mix financing, real estate laws, banking laws, business laws (contracts) and perhaps a builder's insolvency (bankruptcy) and you probably can count the majority of attorneys as non-proficient practitioners.

As I mentioned earlier, there are competent attorneys who specialize and are experienced in these issues, but unless it's a high-profile case, finding them is not going to be easy, if not impossible. Most state bar associations are of no help. In one case, I personally interviewed over 50 law firms (attorneys) in a three-state area never satisfying myself that any one firm had the experience or enough knowledge to handle the construction issues much less the peripheral issues that always seem to pop up. When you consider only the facets of law I've mentioned in the above paragraph, you will realize that you're in for some serious cash outlay. That's if you're lucky enough to find the right law firm for the job. If not, why bother? The general tendency is to engage the cheapest attorney, hoping everything will take care of itself. In the long run you will end up paying more!

Always have a professional, someone experienced in construction matters, work with you and your attorney when drafting contracts. Your construction professional might not know the law, but he will know the business at hand. It's a good idea! *The reality is that most attorneys are technicians, not savvy businessmen, which is a prevailing misconception.* The "feast or famine" mentality, as evidenced to me by the substandard loans I have made over the years to attorneys with poor credit, speaks volumes.

And when it comes to understanding the concepts of negotiation and leverage, forget it. Many seem to want the case to turn on a point of law instead of settling based on leverage their client may hold. That's insanity, non-business sense. All logic goes out the window when you start dealing with attorneys. They have their own agenda. Egos get in the way and case loads burden those who are capable.

Similarly, most judges are inundated with drug cases and seem to be unintelligible about construction issues, business matters, and property title.

Jurors, (butchers, bakers and candlestick makers) for the most part, won't have the slightest idea what you're talking about.

I find the judicial system, for all practical purposes, in and of itself, is an obstacle when it comes to resolving these types of issues.

☞Circumvent the court system by including remedies and cures within your construction contract similar to those outlined herein and make sure that your remedies are thoroughly explored in all their facets. Answering the following simple questions: Who? When? What? How? How much? and What for? can lead to much broader-based, thought-provoking solutions (remedies). Use a capable attorney along with a construction specialist to implement these measures in your original contract.

Your contract must be thorough and practical, enumerating your remedies and their implementation and timing. Otherwise, when the impasse comes, you'll find yourself in an adversarial position spending your valuable time and good money negotiating parameters that should have been in your original contract, ultimately, with results you may not like. During contract negotiations is the time to define parameters to deal with impasses not at the point of impasse.

Two final thoughts! Firstly, lawsuits are costly, and some builders have a tremendous cash-flow generated from ongoing jobs that can be diverted toward their defense. Keep it all in perspective. It's your money. Don't ever loose sight of that. Never allow "the principle of a thing" to fog your thinking. Stay out of court!

Secondly, builders insulate themselves within corporations. If you do come up a lucky winner you may never see any revenues if the builder's corporate veil is declared insolvent. Why throw good money after bad to find out? Again, leave the "principle of the thing" to someone else. Your chances of "nailing" an unscrupulous builder are remote. **Stay out of court!**

5.) Never sign a contract with only an estimated completion date or deadline. Always state the specific date construction shall be completed. Define what construction completion means and what circumstances shall be met for it to be deemed complete. If your builder insists otherwise, he is engaging in inappropriate construction practices.

6.) I wish these pages could talk because I'd quirt them until they screamed, w*atch out for Slick Eddie*. He comes in all sizes and shapes. Husband and wife teams do a bang-up job at this. They'll draw you in with a low-ball cost estimate promising to sign a lump-sum contract, but somehow along the way, they will convince you that they can save you money, if only you would instead, sign a cost-plus contract. There you have it. I've told you again! **DON'T!**

7.) Define in your contract reasons and methods for discharging your builder of his duties, who may discharge him, when and under what circumstances he may be discharged, what will be paid at discharge, and when payment will occur.

I simply state, "upon written notice the builder may be discharged at any time by the owner at the owner's sole discretion." Make sure you define what payment is due and in what timeframe you will pay after his discharge. I would also address damages owed, if any, if that juncture ever arises. You might consider writing damages out of the contract unless your builder misses his construction deadline. But if you ultimatley discharge him for breach of contract and have to hire another, you most likely will incur additional costs. Those costs should be addressed in your contract. For example, interest payments, errors in construction, builder transitional expenses would include new subcontractor contractual costs and delays, etc.

In attorney's jargon, make sure you are made whole again, which simply means, recover your losses.

If discharge is contemplated, it is a good idea to update and fine-tune your bookkeeping. Accounts must be settled quickly. Engage your next builder before executing the discharge; hence, less time will lapse in the construction schedule.

Caution: If your builder carries (originates) the construction loan for the project, it will become a significant hindrance in discharging him. In fact, it may deem it impossible.

Also, if you don't own the lot, you probably won't have the authority to discharge him. Your only alternative may be to walk away. But all options must be addressed in your contract.

Seek professional advice in these matters.

Chapter 10

TALK IS CHEAP!

(Oral Contracts)

...I'm not sure they (husband and wife builder) have the concept of what's fair, and what's legal, and what's right. I think they know exactly where the borderline of gray is, and they live on that borderline of gray. I highly recommend anybody not to use them because it doesn't matter what they say or promise...

WRITING IS BELIEVING: WHAT'S IN AN ORAL CONTRACT?

Consider oral contracts or oral agreements poppycock. Once disputes arise, interpreting written contracts previously agreed upon becomes amazingly difficult. *Document everything in writing maintaining your paper trail.*

Remember, if you become lax and allow changes to be made without written documentation, these actions may eventually work against you. For instance, the builder may cite that you authorized a change verbally, pointing to your verbal authorization of other changes in the past.

If you're under the assumption that your builder is listening, he might not be. If it's not in writing, don't expect to see it in tomorrow's construction. Chances are it won't be. Some builders are infamous for "not" staying on top of paperwork and usually for very good reasons. Once construction starts, work may progress like a runaway train, or the opposite may occur, and you may never find a subcontractor on the job.

I strongly recommend that if you are facing the *runaway train syndrome*, you hire a professional to represent you immediately. On the other hand, if work isn't progressing, find out why. It must be continuous to maintain the construction schedule. Hopefully, you've entered into the project with a complete set of plans, specifications, and a good contract.

Chapter 11

A FIST FULL OF DOLLARS!

(Financing)

...as someone reminded me, it only took 23 months to build the Empire State Building. Why then, did it take my builder over two years to complete my home which is less than 3,000 square feet?

TIT FOR TAT: WHAT'S IN A DRAW? AND ROME WASN'T BUILT IN A DAY!

The longer the term of construction, the more interest you will pay on the construction loan. You may never recoup it. Don't think it won't happen to you. It's happening every day.

I've witnessed a builder take nearly two years to complete construction on a 2,500 square foot semi-custom home with only an estimated date of completion. Construction financing was carried by the owner and the house was built on the owner's lot. The same house was finished as a spec in less than four months. Go figure!

Working without defining a definite completion date, some might say, is like playing Russian Roulette! *Never use an "estimated" date for construction completion. Always insist on a firm construction completion date and a firm price.*

MONEY IS MONEY: DIFFERENT STROKES FOR DIFFERENT FOLKS!

Generally, there are two types of construction financing available: the **one-time close** and **conventional construction financing** (interim loan) completed with a take-out loan or mortgage. Normally, a construction lender will loan up to 80% of the total cost including lot cost.

The **one-time close** was developed to ease the transitional costs from construction financing through to the take-out or the permanent mortgage. Each time a loan is originated, there is a "close". This means a title search must be completed on the subject property to discover any liens, affidavits of completed work must be prepared and signed, and loan paperwork must be drafted for signatures. When originating a new loan, a trustee is established who secures the property for financing. It all takes money, which translates into fees to the consumer. They add up!

Until recently, constructing a home took a combination of loans. A **conventional construction loan** bridged the gap through construction to finished product. At completion, the consumer paid off the construction loan with a permanent loan (mortgage) or take-out loan, all one in the same. This method necessitated two loan closings, two sets of documents; therefore, more fees. The one-time close eliminated the need for two loans, not to mention the need to qualify for but one loan.

You must compare the interest rates and closing costs associated with each type loan to determine which best suits your needs. Rates and fees vary. Normally, the conventional construction note carries a higher interest rate due to risk associated with construction and shortness of term amonng other things, while the permanent mortgage (put in place at completion of construction) reflects a lower note rate. The one-time close, a combination of both, reflects a blend of the note rates, similar to a weighted average, which means over the term of the mortgage you might be paying more.

Until recently, only a limited number of national lending institutions were knowledgeable enough to make the one-time close loan. Kinks now having been resolved; smaller banks are getting into the act.

The good news is that smaller local banks, almost a thing of the past, normally demand fewer fees, but they also may be understaffed, their personnel not highly proficient in construction vernacular.

Therefore, when your builder requests payment, make it your priority to verify that work has actually been completed. Visit the job site! In your original construction contract, elaborate on the procedures of making a draw. Make sure copies of ALL paperwork and invoices are included in the request for draw from your builder. Have them forwarded to you several days to two weeks in advance. You need time to review the paperwork and check the work at the site. Don't be cajoled into anything else, especially if the builder and lender have been adequately forewarned. Realize that paperwork can be complicated, especially when your builder is disorganized by nature or purposefully.

PAY AS YOU GO: WHAT'S IN A SCHEDULE OF VALUES?

A **schedule of values** represents a cost breakdown of the construction by job category. Note, it is not called a schedule of costs. Can you guess why?

The builder prepares it based on the overall cost per the lender's instructions. Each lender predetermines what job subcategories the builder must complete in order for the lender to release advanced payments, draws, from the construction loan. Each category represents a percentage of the total loan.

A typical schedule of values may include the following categories: lot cost, professional fees, appraisal fees, soils test, site grading, site preparation, engineered foundation, slab plumbing, concrete slab, framing, cornice, sheathing, permanent roof, insulation, plumbing rough-in, heating rough-in, electrical rough-in, windows and doors, sheetrock, cabinets, vanities, countertops, exterior finish, siding, plumbing fixtures, trim, tile/bath surround, painting, walks and driveway, landscaping, air compressor, lighting fixtures, floor covering and decks.

WHAT'S IN A DRAW?

A **draw** is a payment made by the lender to the builder for work completed. Not to be confused, while termed an Advance Payment, in actuality, payment is made for work completed based on the schedule of values. See schedule of values above.

Make certain that your original construction contract defines the number of draws the builder will be allowed to make. Usually, four is sufficient.

If a builder is not financially capable, he may ask for more draws. If construction costs are over $500,000 extra draws may be required. Each draw has associated fees attached similiar to a "close"; therefore, the more draws, the more fees you will pay that will never be recouped. *Negotiate and restrict the number of draws* in writing in your original construction contract and don't budge. There is no need.

In advance, ask your banker, builder or, perhaps, secure a copy of a typical draw form from your local A.I.A. office. Becoming familiar with its format is important.

One last caution. Your builder may outline a payment schedule in his construction contract.

Coordinate any draws contemplated with recommendations from your lender before signing any contract.

Chapter 12

FOLLOW THE YELLOW BRICK ROAD AND OTHER MEANDERINGS.

(The Construction Process)

Our son, a recent amputee, was confined to a wheelchair. We informed the builder that everything in the house must accommodate his needs. The builder understood and told us about the awards he had won for handicapped-accessible construction.

We were glad to have found him. We paid him a large down payment so he could get underway with our new home immediately. Our present house would not accommodate our son's special handicap needs.

Construction started, but neither the house nor the site was ever cleaned. It was impossible for us to visit the job site during construction with or without our son. An exterior wheelchair ramp that we had requested was never constructed. At the end of the job, I had to hire someone else to build and install it.

Finally, the builder informed us that construction was complete and that the house was clean. We entered with our son, and we couldn't believe what we found.

The doors weren't wide enough for his wheelchair. The special shower we had requested had not been installed, and his wheelchair wouldn't fit through the shower door. The vanity top in his bathroom was installed too low to accommodate him.

We had paid the builder everything he was owed, and now we were at the end of the job. The builder had drawn all the construction money from the bank. We had no leverage left to force him to comply with what he had promised. We had nothing in writing. What could we do?

The builder wouldn't respond to our request to make things right. In fact, we couldn't even find him. Our pleadings fell on a cold heart and deaf ears.

WHAT PLANET ARE YOU ON & WHO WILL INTERPRET THE ISSUES?

Already discussed above; it's worth repeating. *It would be wise to define in your construction contract that a third party, a professional designated by you, will interpret plans, specifications, scope of the contract, the contract's intentions and whether or not completed construction meets all the requirements laid out in your construction documents. His decision must be final!*

Before signing any contract call a pre-contract meeting at which time you will thoroughly review all documents with the third party, your representative, your builder and any other professionals you deem appropriate. Perhaps an engineer. All parties must comprehend and agree with your understandings and your interpretations of the contract douments. Your builder must be part of this meeting. Otherwise, there will be misinterpretations on his part. Including everyone in a pre-contract meeting will further ensure that the project will go smoothly.

DOWN PAYMENT, SAY WHAT? CHEESE!

Wait just a moment! Some builders are masters at confusing the public, and that's not where the term master builder comes from!

NEVER, NEVER, EVER make a substantial down payment to a builder. To "seal the deal", it's OK to pay a token amount at execution of the contract, but make it understood that it will be deducted from the first draw.

The only other instance when I would even consider an advance payment would be for a job over $650,000 with excessive or expensive custom work. Millwork including windows, cabinets, and casements, which are designed and assembled from scratch, would fall into this category. Items fabricated or shipped from overseas would be another category. This doesn't include job-built or semi-custom cabinetwork on smaller jobs.

A portion of the foregoing custom-work cost may be required upfront. This must be analyzed on a case by case basis and before the construction contract is signed. If it appears reasonable, it may be acceptable to pay the "job-shop" (the one creating the custom work directly at the builder's instructions), but make sure your builder is responsible for any **as-built dimensions** (job site measurements) required for such work. Not only does correctness count, but timely forwarding of as-built dimensions can become critical to completion of your project. Define how and when measurements are to be forwarded to the "job-shop" by your builder and make sure to include your custom work in your *critical path* as explored below. Make your builder responsible and accountable for all custom work; after all, he chose them for the job. Stay out of that loop. Otherwise, there will be a dilemma if work arriving to the job site doesn't conform to specifications. Your builder should make periodic visits to the "job-shop" if necessary to ensure work is progressing according to schedule. It wouldn't hurt if you double checked the progress and quality of work underway, but work through your builder.

If the builder is financially frayed and can't carry you through to the first draw as prescribed in the schedule of values and in your contract, you may be seeing signs of problems ahead. (Construction managers are not builders in this sense.)

For example, I've witnessed jobs shut down without notice, right in the middle of construction. Subs stopped showing up. It was discovered with a little investigation that the builder was in financial disarray and wasn't paying his vendors or his subs. To get the job back on tract, the owner had to pay the subs directly. Once the job restarted, the builder insisted that the owner pay him a mark-up on all work and materials even though he was no longer in the construction loop. This demand was based on the originally executed cost-plus contract.

KNOW THIS!

1.) Never pay for work that has not been completed.
2.) Never pay for material that has not been delivered to the site and is not secured from theft and damage. If materials are damaged or stolen, the builder's insurance should pay for those losses. That is the industry standard. Don't let anyone convince you otherwise. You want to define that in your contract.

Another industry standard which has lost its gusto in residential construction is the retainage, a percentage, usually 5% to 10% retained from every payment drawn.

The retained amount should be paid to the builder at the completion of construction and only after you have accepted the construction as finished. This helps assure completion. Retaining a percentage of the payments is adamantly enforced in commercial construction. But in home construction, due to the lack of educated professionals participating and the quality of builders, it is slowly disappearing. But nonetheless, it is a sound mechanism still used by those in the know. If your plan is to retain a percentage, define it in your construction contract. But whatever agreement you finally negotiate with your builder, never consent to pay him money before work is complete.

ON TIME AND IN BUDGET: A PRIORITY, NOT AN OPTION!

Before executing the construction contract, confer with your builder and develop a time line or schedule of construction. You are interested in the **critical path**, a series of events or jobs that lead to construction completion. The critical path establishes the shortest amount of time necessary to construct your home; therefore, finishing on time is not an option, but a priority. Reference the time line in your original construction contract and attach it as an addendum. As construction progresses, time adjustments made by change orders may, but not necessarily, extend the completion date. The time frame should remain accurate within a couple of weeks.

The term critical path is used more often in commercial construction, and you might find that your builder is unfamiliar with the concept. That doesn't make it any less valuable.

The critical path is the *shortest path* between commencement of construction and completion. Therefore, the first thing to remember is to define in your contract when construction will commence and what conditions deem it complete.

On this path, certain jobs, *critical jobs*, must be completed before other jobs can start.

It logically follows that the job site must be cleared and treated for termites before the plumbing is dug and before the slab can be formed and poured. Wall framing is then erected on the slab; next comes the roof and exterior sheathing. The inner workings of electrical, plumbing, and

HVAC (heating ventilation and air conditioning) are then placed within the walls, floors, and ceilings.

Sheetrock covers the inner workings within the walls and hangs on the wall framing forming a surface on which to finish. If a **critical job**, that job or stage that must be completed (or started) before other jobs can commence is interrupted, all other jobs that follow will fall behind schedule until that critical job can be started or completed.

To expand further, in the example above, you understood that clearing the site and termite treatment are critical jobs that must occur before the slab can be poured. The slab becomes critical in order for the wall framing to be erected. All electrical, plumbing, and air conditioning must be placed within the wall framing before the sheetrock can be applied to the wood studs and ceiling framing. The critical path (jobs) now becomes obvious once you stop to think about it. It's not rocket science. Finally, multiple jobs begin to occur simultaneously, but if you look closely, there are other jobs that must proceed ahead of them, as we've just mentioned.

To keep on track, establish the critical path with the builder and question him if it is interrupted. Make it a part of the contract. The path simplifies the construction process enabling anyone to stay on top of the progress. Crosscheck the critical path scheduling with your inspector or representative, or better yet, include him in developing it.

The critical path may be represented as a bar graph. Graph paper will make it easier to sketch. Your lender's "schedule of values" may help you determine the sequence of work. (See Appendix A)

For the first job, start with day one and draw a horizontal line from left to right representing a segment of the total time to complete. The next job extends to the right from there or overlaps. Each line drawn represents the length of time estimated or scheduled for each job. For instance, site demolition (tree removal, etc.). At the end of that line, extend another line, maybe with another color pen representing the time estimated to prepare (grade) the lot before forms can be erected for the slab. Next, extend a line that represents the estimated time to set the slab forms.

Continue the bar graph to the right with successive jobs until you reach the date construction is to be complete, the completion date. Note, once you move past the framing stage as discussed above, multiple jobs will begin to occur simultaneously. Don't panic! This is natural and must happen if the house construction ever has a chance of finishing within a reasonable time-frame.

Just extend multiple colored lines stacked on each other. Name them according to the job, representing each job within that timeframe. The length of line represents the duration of that job. If you like puzzles, this can be a pleasurable experience. It'll really get you into the project. The critical path will soon be visible.

Find and follow the continuous progression of jobs, jobs that <u>must</u> be started or completed from start to finish.

BRANDS: IDENTIFYING MARKS OR WHAT'S IN A LABEL?

It never fails. I can specify the most expensive stainless steel kitchen sink money can buy, and the builder will try to slip in a lightweight builder grade. I get tired of having it yanked out!

If you lack product knowledge concerning items specified for your home, below you will learn how to take control and make the process work for you.

First, in your original construction contract, insist that all *labels and identifying marks* be left on materials, appliances, and equipment until you have had a chance to inspect and verify that they represent what is specified. Not to worry, everyone on the job site should be much too busy to remove labels. Cleaning the job site and removing labels is something that normally has to be encouraged even if written into your contract. Check the labels against your specifications, floor-plan schedules, and contract. Many items are labeled including siding, windows, doors, fixtures, equipment, floor tile, cabinets, and countertops, etc. It never hurts to double check.

Another available paper trail is the *vendor delivery ticket*. These should indicate which items were delivered, by whom, and who accepted and signed for them at the job site. If you are not sure whether an item installed was specified, refer to your cut sheets to verify it as the correct item or take a picture of the item and bring it to the vendor to authenticate. Does it match the item that was specified? In your contract, state that you must receive a copy of the delivery tickets before payment is made.

Insist that all specified items be installed. Allow no substitutions, which means, you don't want credits for substituted items of lesser quality without your prior written authorization. It pays to stay on top of things.

YOU'VE GOT YOUR MARCHING ORDERS: WHAT'S IN A FIELD ORDER AND A CHANGE ORDER? HOW ARE THEY USED?

Normally, a change to scheduled work is requested in the form of a **proposal**. For instance, you would ask your builder for a written proposal to change your square front door to a round door. His proposal may come back to you in the form of a *change order*.

A change to the work made after construction commences may increase the construction cost and may extend the completion date if you agree to it.

Paper documents that record work changes are *field orders* and *change orders*. These documents should be considered additional contractual agreements altering your original contract and should always be in writing attached as part of the original construction contract as they are executed.

The **field order** is normally used to record minor changes made in the field during construction. Usually, there are no extra costs involved and the completion date is not affected due to the change. A field order should be in writing. The **change order,** on the other hand, is used when there are changes that affect costs and/or construction completion issues.

Authorized field orders or change orders must be in writing. Define that in your contract. Also, insist that you must approve, sign-off in writing, any change before it occurs on the job site.

Unscrupulous builders abuse the change order to increase costs. Abuse occurs primarily due to a loose set of plans and specifications. If nothing is defined or specified in your construction documents, the builder's imagination may start running rampant. Every item you choose becomes an additional cost. Misuse of change orders can wreck any construction budget.

Again, change orders were developed to record changes in construction. A change order documents changes in work, including all costs to make them and any additional construction time required, as well as, any credit for work not completed due to the change. *You must insist that changes not commence until all the paperwork is completed, and you have agreed to the price and any additional construction completion time.*

Insisting that paperwork be completed and authorized before ANY change commences allows you time to review before making your final decision. It also allows you the opportunity to reconsider additional construction time requests from your builder. Of course, just because you asked for a proposal, doesn't mean you have to authorize (sign) the change. If you are not satisfied with the pricing or wish to reconsider for other reasons, now is the time. You don't have to follow through.

A change request (in the trade called a "request for proposal" or **RFP)** may incur additional costs, whether warranted or not. In your original contract for construction, in writing, *define when additional costs will be due the builder if there is an authorized change.* If you don't, you'll find yourself running to the job site with a check or cash in hand before any work will commence instead of being able to pay for additions or changes in an orderly manner within the normally prescribed course as set out in the draw schedule.

It's a good idea to specify in the original contract that all paperwork concerning additional work shall be prepared by the builder and authorized in writing by the owner in a standard A.I.A. change order format before commencement of such work. Any payment for completed additional work shall be due at the next scheduled draw per the contract (construction contract).

Contractors would like you to pay up front for the changes. Don't do it unless, of course, it is a speculative home in their inventory. Then, you may have to pay for any changes or additions up front, because your personal choices may be contrary to or in addition to the builder's original plan. That will present a dilemma for you.

The builder in this instance doesn't want to be responsible for the cost, especially if you void the contract.

Why? The reasons are numerous. You may be over-improving for the neighborhood, or your selections may not be considered mainstream by the builder. Realize that on a speculative or a turn-key project, you might not get your money back on changes made if you default or void the contract unless otherwise addressed in your original contract.

Also, you must address in your original contract what mark-up charges (profits & overhead costs) the builder will be allowed to make or must deduct on labor and material changes. This will help prevent price gouging. This is true for any style contract, whether it be lump-sum, cost-plus, or any combination thereof.

Using a **hemorrhaging change order** technique, unscrupulous builders escalate original construction costs to unimaginable heights. When an original price quote of $187,000 ends up costing $265,000, something is wrong. That represents a 42% increase, no matter how you slice it. Again, this is primarily due to loose construction documentation (plans, specifications, and contract). It breaks the bank! Caution: additional costs may have to be paid in cash by you, the owner, if the lender doesn't agree to include them in your line of credit or your mortgage.

You've been forewarned!

One last item concerning changes. Never approach a subcontractor and ask him to make a change. Always go through your builder or his superintendent. If a subcontractor implements a change for you without the builder's authorization, no matter how innocent the change, the results could be disastrous. It could end up costing you far more than you could ever imagine, especially, if another subcontractor's work is damaged during the process.

Chapter 13

SEEDS THEY ARE A SOWING…

(Jobsite Issues And Accountability)

…what he says, ain't necessarily so. He was born to lie.

Unfortunately, he convinced us to use a cost-plus contract. He said he could save us some money. We contracted to pay him 17.5% for profit and overhead.

I have an original invoice from the cabinet company. We had no idea the builder owned it. The builder told them to add an additional 15% for him to the price of the cabinets that we bought. In other words, what we paid was 32.5% profit for the price of the cabinets instead of the 17.5% that we had contracted to pay him.

When we contracted with the builder to build the house, we told him to build the house at the lot elevation and not to lower the level because of the hurricane threat. We were seventeen and a half feet above flood stage. He called me in Vermont at our summer home. This was after I had signed the contract. He told me that his subcontractor didn't like the smell of the soil and that he would like to get someone out to get a soil sample to see if we needed pilings driven. I told him that since we were building our dream home, we didn't want any problems or complications at the end. He called back and told me that the soil samples were done and that we needed to drive pilings to support the house, which I agreed to do over the phone. Sometime later in April when we arrived back on the Coast, I asked him for the soil report. He told me he had it at his office, but he never produced it.

I had asked him over the phone what the normal thickness of the slab was? He said four inches. I said instead of doing it four, since we were going through the expense of driving pilings, let's pour five inches.

Unfortunately, in his deceitful way since he knew it was a cost-plus job, instead of pouring five inches, he came in and just started pouring concrete. In doing that, the plumbing collapsed under the weight of the slab. Later, in fixing the plumbing, we found that the thickness of the slab was anywhere from nine to fourteen inches thick, which meant additional cost to me along with the 17.5% mark-up.

In the meantime, the builder had to jackhammer the concrete slab from the master bedroom bath all the way to the front dining room, which is approximately sixty feet of slab. It had to be jackhammered to fix the plumbing, which damaged the integrity of the slab. The floors were finished in expensive tile which also had to be broken out. I paid for that and also to replace it.

The new slab was jackhammered an additional four other times because of different problems including: not putting in the drain for the icemaker, which was included in the plans; the toilet didn't work in the master bedroom because the plumbing had collapsed, and the range draft for the stove that ran in the slab should have been six inches instead of four. The plumbing in the kitchen was also offset. The reason for that, according to the framer, was that the builder gave him plans for a stick (wood frame) house and not the concrete I.C.F. unit house for which we had contracted. It has thicker walls. That offset everything, and that was a major problem. We ended up paying for all the builder's mistakes. The guy's a con artist.

BLAME IT ON RIO: WHOSE PROBLEM IS IT ANYWAY?

We've covered a lot of issues, and, hopefully you will have smooth sailing from here, but chances are good, if you're involved in construction, you will encounter other problems. Remember, it's not brain surgery. Most of the laborers on the job learned their craft from doing just what they are doing. Often a builder will learn from the subcontractors, or maybe not!

The following list is offered as a heads up. Be on the lookout for the following conduct. I have also offered some suggestions on coping with such actions.

1.) The warranty period begins after you, the owner,

"accept the construction" effectively putting your builder out of the house. Unscrupulous builders, who don't honor their warranties, not even their one-year warranties, are on the rise. They seem to be affected by a lackadaisical warranty work ethic. They play a shell game, putting off warranty work until either you go away or the *warranty period* expires. First, define in your contract when "exactly" the warranty period begins. It should begin at the point construction is completed and accepted by you.

One way to stop this unethical lackadaisical warranty work abuse is to define in your original contract options available to you as a homeowner in the event warranty work is not completed within a prescribed time period, say within a week of notifying the builder in writing, two weeks tops. One option may be to have the work completed by another contractor at the builder's expense. Make sure that your builder understands and acknowledges in your original contract that under these extreme circumstances his warranty shall still remain in effect even though you've been forced to use extreme remedies. Also, spell out in your original contract that this clause survives construction completion and acceptance of the house.

2.) Another serious ongoing problem is *material not showing up on the construction job site,* it having been diverted elsewhere with invoices continuing to roll in.

Well, you say this can't be happening in today's world. This is no joking matter. I've had the unpleasant experience of knowing a builder who constructed his million-dollar home from materials he robbed off other construction projects he had under contruction.

This has to be discovered timely, if not, you'll never catch up. First, in order to detect this abuse, you'll have to have an original estimate of the quantity of materials, preferably prepared by a third party. If you're working under a lump-sum contract, this type of abuse will not affect your pocketbook, unless the builder becomes insolvent. If you're working under a cost-plus contract without a firm price or ceiling, you've got major problems.

One way to stop this, if you don't have a full-time site representative, is to state in your original contract that you will not pay for materials that have not been accepted and signed for by a mutually designated on-site representative. Upon site delivery, a copy of the delivery ticket must be forwarded to you immediately.

Now, if you suspect a problem, you have the burden of cross-checking the delivery tickets with the vendors and their drivers to verify what materials were actually delivered to your site. Involve management and they will work with you.

3.) Another serious problem is *waste of materials*. I've been on jobs where a framer cut a two-foot square from a 4 X 8 sheet of plywood and then threw the sheet remainder into the dumpster. Can you guess what style contract they were working under? Again, if you're working from a cost-plus contract, you've bought the farm. Under a lump-sum contract, you are covered, even though I hate to see waste, especially when it comes from natural resources such as trees!

4.) All too oftern, I've witnessed subcontractors, especially plumbers and electricians, obliterating the wood framing, the structure of the house. Once the wood drills "power up," it takes only minutes to literally *destroy the structural integrity of wood framing*. I've run electricians off the job for less. Instruct your designer to address maximum allowable notching and cutting in his plans and specifications, not only size of notches and holes, but quantities and location within structural members.

5.) Masons who are not knowledgeable or who don't

care allow the cavities between the brick veneer and the weatherboard to be filled with mortar and even throw pieces of brick into the cavities, thus leaving less clean-up for themselves at job's end. This *destroys the integrity of the wall* system effectively creating a dam within the cavity. Water can't properly drain causing moisture penetration within the interior of the house. A good detailed set of plans is your best safeguard in this instance, but if you catch this abuse on the job site, you've got to stop it. It's not cost you're worried about; it's the structural integrity of the wall that can be affected, not to mention mold and mildew problems.

6.) Make sure there are holes into the attic behind vent screens!

7.) Another problem I've encountered is *work stoppage*. Rain delays which may stop construction begin to smell fishy once the roof is in place. The whole issue of rain delays should become moot after construction is dried in (roof on, windows and building wrap in place, doors shut), but you might be surprised.

Delays can signal a myriad of problems. Builder's cash flow problems may hinder his ability to pay the subs, or he might have pulled crews off your job to finish or start another job. In any event, work stoppage warrants further investigation and must be remedied as soon as possible, because it interrupts the critical path. You, like others, might be forced to pay subs directly to get them back to work. This is one good reason to define damages due to missed construction deadlines.

8.) Another issue that may surface involves the selection of materials and equipment. In your original construction contract, make sure that all options are available to you for choosing materials and equipment. Don't leave this to chance. Spell it out in your original construction contract. *Don't allow the builder to limit vendor selections*, especially on elected changes. Make sure this is in writing. In fact, define how selections are to be made and in what timeframe. If the builder directs you toward a vendor, it could cost you additional money. If you've found a product you just have to substitute for an original selection, insist that the builder install it, but first get your books in order. Use the change-order form (defined in your contract). In writing, have your builder credit you for the cost of the original product or item as specified and the installation cost associated with that item. Then, insist that your builder document in writing additional costs, if any, to be incurred from installation of the new product. Use common sense. Installation costs are already in your original budget for the original product. Do the new selections justify additional costs to install? If so, why?

A prime example might be the substitution of a front door. Your wife stumbles upon a front door that she can't live without. You've checked the job site. The original front door has not been delivered to the site. The new door selection costs approximately $1,800. The cost appears reasonable; therefore, a decision is made to make the change, but the builder comes up with a lame excuse explaining that he just can't install that door. No logical explanation is given. He insists that a selection be made from his vendor catalogue. The same door ends up costing you $3,700. Watch out!

Based on what I know about the incident after interviewing the owner, this builder was probably receiving a kick-back. Attorneys advise me, that under the right circumstances, these acts fall under the R.I.C.O. act, racketeering, which is a criminal offense. That means jail time.

9.) Watch out for the *runaway train syndrome*. After discussing proposed changes with your builder, you note that he allows construction to continue before he presents you with change-order paperwork. He isn't stopping or slowing construction allowing the orderly process of the change to queue into the sequence of construction. Instead, he ignores your request and continues to steam ahead.

Later, he charges you for tearing out the construction, that which he has negligently allowed to continue. In other words, he hasn't scheduled or included your change in the orderly process of work. Now, in order to accommodate your change, he will claim you are responsible for his lack of management and scheduling. This is a serious problem and probably only a CONFIDENT professional will be able to help if this continues to occur. Again, remember in your original construction contract to define how requested changes will be scheduled for sequencing work into the process. WATCH for this, especially if you are working from a cost-plus contract. You'll pay for his mismanagement and mistakes, unless you contract around it. This type of behavior is significantly damaging under any contract.

One safeguard is to take pictures of the area where the requested change will occur. This will document the stage of construction completion on the day you requested the change. Have the picture date stamped and make multiple copies. Most one-hour photo drops can accommodate you with quick picture development. If you anticipate problems, deliver or send a copy with a letter A.S.A.P. to the builder summarizing and documenting your request for the change. This effectively creates a time frame of events establishing for all concerned what, when, and how events sequenced. But realistically, if the relationship has deteriorated to this point, it's time to call in the professionals.

10.) Another unethical practice is *bait-and-switch.*

Builders take unsuspecting clients to a magnificent model home informing them verbally that everything they see is standard in all their homes. But, once the ink is dry on the contract, the builder develops a terrible case of CRS (can't remember s…), conveniently forgetting all his promises. What was previewed, AIN'T what he's building. Perhaps, now, a new buyer occupies the model, and you no longer have access!

Your safeguards will already be in place if you've followed the recommendations in this book. Document everything in writing with photographs if necessary, and attach and reference them in your construction contract, with copies saved.

11.) Be cautious of the builder who comes to you during construction, especially if you are working with a loose set of plans, and asks you to make your selections from several items he presents to you at that time. Before you know it, he has installed your choice and has also "up charged" (additional cost) you, when all along *you thought the item selected was within the original budget.* Again, no additional charges should be allowed unless you have authorized them in writing. Write it into your contract.

12.) Fact #1: Successful builders may be astute businessmen. Fact #2: In most states a person with an insignificant financial statement has to do little more than fill out a form and figure out which end of a hammer to pick up to call himself a builder. Fact #3: An astute builder will cut costs to improve profits, and his cost cutting starts with skilled labor.

Therefore, insist that your design professional include instructional phrases such as, "All work shall proceed in a workmanlike manner, be level, plumb and square with tight joints and corners..." Your designer knows the drill.

13.) If you plan to supply material or other items,

you must address additional issues including but not limited to the following: Who is responsible for damages to those items? Who is libel for improper installation of the item(s)? Who is libel for damage to items that have to be torn out in order to replace improperly installed items? Who is responsible for replacing such items? These issues become very complex.

14.) Define in your original construction contract circumstances whereby you may discharge your builder. You must be able to discharge him with written notice. The issue of what is owed upon discharge must be addressed. Items satisfactorily installed at the job site or stored (protected from theft, weather, or other damage) properly at the job site and not damaged must be paid for, but not until the next draw following discharge. If necessary, address the establishment of an escrow account to pay for such items.

Define in your original contract that discharge may occur for breach of contract and under these circumstances any additional charges or fees incurred due to the transition to your next builder are considered your original builder's obligations, and may be deducted from any sum owed to him.

15.) In your original contract, deny your builder the right to lien or lis pendens your property, whether he is discharged or not.

16.) Watch out for the *forget-me-not syndrome.* It starts innocently enough with little things the builder just can't seem to remember. Before long, you find yourself on the job site full-time making sure work is properly completed according to plans. You're, in effect, running the job.

Suddenly, this forgetfulness the builder developed escalates into the substitution of lesser quality items. He may also start ignoring your change order requests. *Stop him dead in his tracks.* If you can't take or don't want the hassle, its time to hire someone who can do your bidding. Your construction documents; plans, specifications, and construction contract will protect you in these situations, but you've got to make him do it. Be firm!

17.) Make sure your framer includes nailers the length of every corner for securing sheetrock, including but not limited to locations where ceilings meet the walls. It's important.

18.) A last bit of wisdom concerning the issues discussed in this book. If you don't feel comfortable addressing these issues with your builder, tell him your lender wants them in the contract and then have your lender shoot him a letter to that effect! It might seem overly cautious to you, but it's only good business.

LIFE'S WHAT'S HAPPENING WHILE YOU'RE DOING OTHER THINGS!

(Insurance Issues)

The fire destroyed everything...it wasn't covered in my contract.

Before he ordered the materials, the builder insisted that I pay him in cash for all the changes we made. We never questioned it.

Now, he's telling me that his insurance probably won't cover the $48,000 worth of upgrades we made. I paid him in cash, and there is no documentation showing that the upgrades were in addition to the original construction cost, or that they were even installed.

What am I going to do?

WHO'S LOOKING OUT FOR YOU: INSURANCE INFORMATION ON A NEED TO KNOW BASIS.

Have your builder include you as a co-insured in his insurance policies. The builder must carry at least workman's compensation, builder's risk (fire, theft and damage), and a good liability policy. In your original construction contract, make commencement of work contingent upon you having a copy of his insurance policies highlighting you as co-insured. Also, you must to be notified by his agent if you are dropped from the policy, or the policy goes delinquent. Make sure the builder is carrying enough coverage. If you own the lot, you might consider including the construction under your present homeowner umbrella liability policy. Check with your insurance agent for further details.

Chapter 15

STACKS AND STACKS OF LETTERS: HOW MANY MORE INVOICES ARE THERE?

(Bookkeeping)

At one point in time, I pulled the landscaping away from the builder and contracted him directly. As the builder drove up one day the landscaper muttered to me, "Oh no!'

When asked, "What's the matter?"

He said, "I'll tell you later."

After I had meet with the builder, the landscaper came to me and told me his story. He wouldn't give me the name of the homeowner, but he assured me that the builder had built a house for a particular doctor who had insisted on a metal roof.

Unfortunately, the builder had given the roofer the wrong set of plans, and the roof was ordered incorrectly and came to the job site precut about a foot too short.

As the landscaper was finishing his contract on that job the builder approached him about the roofing problem.

The builder told the landscaper that he was out of pocket $8,000 to $10,000 because of the roof problem, and that he wanted him to embed his loss in his ivoice, the $8,000 to $10,000. He explained that the owner of the house was a doctor and would never know the difference.

That was their problem, I told him. I had my own problems. Unfortunately, I had signed a contract with the same builder. It called for four draws, a slab draw, a framing draw, a trim draw, and a final draw. The builder sent me a letter sometime in January telling me that, if I didn't pay him a trim draw of about $60,000, he would stop construction on the house.

At that time, I contacted an attorney, and we sent a letter to the builder stating that our contract called for four draws not five, and he had already been paid for the trim draw. Also, in his first draw, *the slab draw, I guess the builder was short of money. He over-billed me about $7,000 on that first draw, and I didn't catch it until after he had been paid. I brought it to his attention. He told me it was an accounting error and that there would be a credit on the next draw. I told him I didn't want a credit on the next draw. I wanted him to cut me a check for the overdrawn amount, but he ignored me. I told the builder that I wasn't pleased, but I never got a check.*

Unfortunately, the first time I got a look at the slab draw application was when we were sitting in the attorney's office ready to make the payment. I was never given an advance invoice to review. All the invoices were presented to me balled up in a paper bag at the attorney's office. They were in complete disarray. There wasn't adequate time to review them. So, I paid, but once I returned home and started scrutinizing them I discovered that the builder had overcharged me about seven thousand dollars for one part of the slab work. I had to wait for my money. The builder ignored my request for a refund check. I had to deduct it from the next draw.

A fact most people don't realize or intentionally push to the background, boring, is that much of the owner's participation in construction centers on paperwork. Staying on top of your job means staying ahead of the paperwork. *Document everything!* Lay a trail of paperwork in order to know what was said, and done, and for what reasons. Besides keeping you updated, it will ultimately help you determine the real cost of your home. *Keep copies of everything.*

If you downgrade or upgrade an item, have the builder document in writing the cost of the original item including installation cost. Then, have him document in writing the cost of the new product including installation. That way you can establish your *credit or debit.* Again, your schedule of values may be helpful here. If necessary, check with local vendors for pricing.

Hire a competent, confident architect, engineer, or other professional in the home building industry to check your job on a routine basis. You'll get a heads-up as to progress. These professionals can also help you with your critical path. If you elect not to hire a full-time representative, I suggest approaching several banks in order to find an independent inspector who will work with you. They are out there. Interview several and check their references. The small amount of time spent on a background check will be well worth it. Make the builder aware of your inspector or representative. It will help keep him honest.

Chapter 16

KNOCKOUT PUNCHES

(Walk-through And Warranties)

My builder is hard to track down. He now addresses out of a P.O. Box. I've seen and talked to him on several occasions at the local building supply, but all I get is, "Yah! Yah! We'll take care of it." Or he acts super stupid, like when I mentioned things such as the water leak. He acts like, what is a water leak? Like he doesn't seem to understand the words, water is leaking.

SUSTAINING OR SUBSTANTIAL COMPLETION—THAT IS THE QUESTION: THIS JOB AIN'T FINISHED!

Substantial completion defines a condition of work that is generally recognized and accepted in the industry as completion of construction, but what does that really mean to you? NOTHING!

It is supposed to *signal* to the owner that the work is complete and ready for his acceptance and occupancy. At this time, the builder also applies to the building official for a *certificate of occupancy,* which certifies that the code official is satisfied that the work is complete to his satisfaction and ready for occupancy.

Substantial completion is a buzz phrase in the industry with which your builder is very familiar, but remember, neither substantial completion nor the certificate of occupancy represents the true litmus test in determining whether construction is complete according to your contract.

This is important to remember. Let the builder engage the code official and gain a certificate of occupancy. What's important is that the builder has to finish the work according to your plans, specifications and contract before the work is deemed complete. Whether construction is complete to your satisfaction or deemed contractually complete in compliance with all the construction documents, has yet to be determined by you and your representative. Remember, the code official is only interested in meeting minimum code requirements that may have little to do with the plans, specs, and contractual obligations of your builder. The builder may try to CON you into thinking otherwise.

When the builder deems his work complete, he will request that you walk the property to verify that fact; thereby, accepting it. Any items that remain incomplete, which should be few, must be listed on a **punch list** for completion. If you begin your walk and start to uncover many items, STOP. Inform the builder that the house is not substantially complete and not ready for punch out. Tell him why. When and only when the house is complete and cleaned should you agree to walk it and list items to be corrected; otherwise, the process is useless.

When I walk a house with an owner, I insist a builder's representative be present. He should bring bright colored press-on labels to be used as visual markers to highlight areas or items that need attention. First, I'll make a quick pass through the entire house. Only after I'm satisfied the house is complete according to the construction documents, will I begin my punch list.

A camera, stick-ons, a flashlight to detect paint overspray, together with pad and pen usually gets the job done. Marking the location of items makes it easier for the subs to find. Again, if I have made it sound like you should expect a multitude of items, that's not the case.

Define in your contract specifically what substantial completion means in your case, but more important, define what the builder will have to achieve before the job is complete according to the contract. Also, give him a timeframe to finish, correct the items on the punch list, and define this as being within his prescribed construction completion deadline. Damages and/or expenses due, if he exceeds his deadline to finish, must be defined in your contract.

Include in your damages and expenses additional interest paid on the construction loan. Whether or not you carry the loan makes no difference if you have to pay man hours necessary to unload and reload furniture, house-wares loaded on and off a moving van, storage cost, and temporary housing cost. These costs will serve as a good start in helping you to determine out-of-pocket expenses. All costs must be defined. Those costs can be substantial depending on time overruns. A lengthy motel stay is not out of the ordinary when completion time is not a concern for the builder. It's your money. Now's the time to "catastrophize" and itemize in detail for contract consideration all expenses you may incur due to construction time overruns, especially if you're moving from out of town or selling your present home.

SHOW ME THE BAG: WHAT'S IN A PUNCH LIST AND HOW TO USE IT?

...it all came down to this...shortly after we moved in, having made provisions in our garage for a large freezer chest, my husband and I started loading it up. I could see good memories overwhelm him as he helped, handing me his prize catches of game and fish. We also managed to get in a custom-cut side of beef and some home-grown vegetables. Later in the week when planning dinner, I looked in the freezer and found that all the food had defrosted and the paper wrappers were now swimming in what appeared to be a pool of sludge.

Upon examination, we found that the wall plug had not been properly activated. We estimated it cost us about $1,500 in food, not to mention the inconvenience of having to clean up the defrosted remains and having the wall ripped out so the plug could be properly installed. The builder wouldn't step up to the plate and cover our cost. So, we literally bought the farm, again.

It's now time to play hard ball. Another good reason to call it a *punch list.* At the end of the job the builder will ask you to complete a **walk through**, walk the home and list items that need to be fixed or haven't been finished in compliance with the construction documents.

FINISHING A HOUSE MAY BE THE TOUGHEST JOB A BUILDER WILL FACE, NEXT TO GETTING YOU ON CONTRACT.

At *close out,* he has to contend with a lot of finger pointing between his subs. During the progress of the job, he has made a substantial amount of money, and if he worked it to his benefit and if you haven't been vigilant, he has very little left to draw from the bank.

Builders are notorious about finding excuses why the work is incomplete. They may eventually end up pointing a finger at you, as if you're the problem. Don't buy it. It's all part of the process of forcing the weakest link to give in. Stick to your guns.

Now, you might begin to realize why a percentage of the job cost is normally held back from each draw. Job completion is the reason! In order to receive final payment, which includes the money withheld to this point and the last draw, the builder must complete the work.

It's a good idea to firm your total cost with the builder before you negotiate the retainage. Unethical or unorganized builders don't like retainage. I've seen some try to stick the retainage as extra on top of the total costs which defeats the whole purpose. .

If there is no retainer, holdback, you may become the *weak link* because he has very little incentive to finish the house. A stand off, with no retainer, will probably mean the builder wins. You will move in, and the items listed on your punch list may never be completed. But if you choose to move in with items incomplete, plan the next six months to be at the beck and call of subs who don't show. When they do, plan on cleaning up after them.

Whether you have elected to retain a percentage or not, *you must hold back the last substantial draw* in order to motivate the builder to finish. Stipulate this action in your original construction contract, "last payment (draw) due upon job completion".

Savvy owners have given builders no quarter and in-turn *builders have become PROS at bleeding construction loans dry, leaving NO money* on the preverbal table before signaling substantial completion or construction completion. It's been a tough lesson, but they've learned it well.

Being lax at this point won't give you a leg up. It will only put you deeper into the hole. Don't give in and don't release the money until the work is complete.

It's now time to get your inspector(s) involved. Have an inspector check the electrical, plug by plug and appliance by appliance. Have a professional test the plumbing drain system and plumbing fixtures. Get an independent contractor to check the heating and cooling systems. If you have hired an independent inspector to help you through construction, have him help you inspect. *It's time to turn over every leaf.* State in your contract, before final payment is made, that your builder shall produce all warranties and instructions for appliances, equipment, roof, including your garage door openers, etc.

Chapter 17

COMPLETION, NOT FOR THE FAINT OF HEART!

(Construction Completion)

We thought it was finally over. Our day had finally come. Construction was winding down, and it would be over soon. The reality turned out to be quite different. It reminded me of bringing my car in for repair. When I got it back, there seemed to be more things wrong then when I brought it in.

Finishing the house was very similiar. For instance, we asked that the light fixture be centered in the breakfast area. When the electrician finished, the ceiling was screwed up. It had a hole in it. It needed patching, sanding, and painting. Some of the tile on the floor had been chipped when he had inadvertently dropped the light fixture. The kitchen cabinets were soiled in dust and mud from the sheetrock repair on the ceiling. All this took a lot of time. A lot of different trades had to come through including a re-clean.

That's just one instance. It seemed to go on and on.

WHO KNOWS WHAT ABOUT ANYTHING: HOW TO TELL IF CONSTRUCTION IS COMPLETE.

Defining construction completion is essential to your success. It may be defined in many ways. Some define it as being the date of *substantial completion*, which arbitrarily can be assigned by your builder. If the contract you are reviewing is unfamiliar, make sure you understand how substantial completion is defined in that contract and make adjustments to suit your needs.

Remember, your rights are up for grabs.

In your original construction contract, state that

construction shall be completed as of a specific date and that date can only be extended by authorized change orders. Insist that construction completion is only achieved when the following conditions have been met:

1.) When construction is completed according to the construction documents. No walk-through shall commence until substantial completion is established, and construction is complete according to plans, specifications and contracted obligations. Substantial completion is reached when and only when contractual obligations are fulfilled including but not limited to the following items:

a.) When all items on the punch list are completed to your satisfaction (or representatives). b.) After all public utilities are connected and turned on and are in your name. c.) After all appliances and equipment have been tested and found to be in proper working order. d.) When final grade is attained according to plans and specs and all landscaping, fences, and ancillary buildings are complete with lawns edged, cut and manicured. e.) After the "final clean" has been completed and accepted. You must define "final clean," i.e., items that must be cleaned and to what extent they must be cleaned including but not limited to: windows scraped and washed, floors, ceilings, walls, doors, hardware, fixtures, appliances, mirrors, cabinets and countertops, and flatwork (paving) cleaned to your satisfaction, labels removed, etc. f.) When and only when all warranties and appliance information is turned over to you. g.) When all material not used in construction has been removed, including but not limited to: junk, contractor's equipment, extraneous debris, including the dumpster and temporary potty. h.) When you have in your possession all material including extra tile and matching paint(s) for replacement or maintenance purposes as specified in your construction documents. i.) When you have in your possession all keys to the property, and the property has been master keyed by a locksmith of your choosing.

Then and only then is construction complete. Don't occupy the house until those items have been deemed satisfied.

But also state that in the event you decide to occupy the house before construction is complete, a reasonable amount of money shall be escrowed, as determined by a third party, until such time as all items on your punch list are complete. That amount shall not only cover the cost of completing and fixing the items, but must also include an additional substantial amount of the remaining money due to encourage the builder to complete construction. Define that substantial amount in your construction contract. It might be defined as a multiple of the actual cost to complete the punch list, perhaps three and one half times the cost to complete.

Also, define the date damages shall commence. That date should be the date construction should have been completed as specified in your contract and may have only been adjusted by change orders. This truly creates urgency for your builder to finish.

Substantial completion is a fairly nebulous term unless its specifically defined or professionals are on the job. If you choose the date of construction completion as being the date of substantial completion without defining exactly what it means, the builder might never finish to your satisfaction. Use the outline above to help you define the construction completion date and be extremely careful how substantial completion is defined and used in any contract you sign. Make sure it doesn't conflict with <u>your</u> definition of construction completion.

Chapter 18

IT'S A WRAP: A GREAT EXPERIENCE OR GET OUT OF MY FACE YOU DIRTY ROTTEN SCOUNDREL!

(Lock-out And Move-in)

I had retrieved all the keys from the builder.

Now, having to be at home to unlock the door for warranty work had become a nuisance. The subcontractors could never come when it was convenient for me. The latest was a repair needed around the chimney. Water was leaking in. I had given up on getting it fixed. The builder's sub had canceled or not shown on several occasions after making arrangements to meet me.

One morning, I decided to take off from work and just relax. My husband was out of town, and I decided to take a late leisurely shower.

After I finished, I walked into the family room as I toweled off. I was startled by the presence of three men standing around the fireplace. I screamed as I covered myself. Thank God, they were as shocked as I. When I finally gained my composure, I found that the builder had given them a key to my house. A key he wasn't supposed to have.

I screamed! I had not been informed about them coming out to fix the leak. Where did they get a key? We were assured that we possessed all the keys. I was so mad, I cried.

I went out immediately and had all the locks re-keyed.

If you don't want it used as leverage against you, never inform the builder of your moving schedule, especially if you don't have a drop-dead date for construction completion. What he most wants to hear nearing the end of the job is, "Our furniture is on the truck." At that point, he will try to drive a wedge. He has been longing for that phrase. Waiting! Strategizing! If you have the option,

I recommend keeping your moving plans to yourself and continuing to monitor the progress of the construction before firming up any moving plans. Anything else will be self-defeating.

Determining if construction is complete becomes exceedingly difficult if you move in before accepting it as complete.

Lock-out occurs before you make final payment, you "square" with the builder. You accept the house construction and prepare for "closing". Finally, the builder is out, locked out of your house. You are now in possession of the keys and control entry. It would be prudent to have the house re-keyed by a locksmith of your own choosing independent of the builder. This eliminates any chance of lost or retained keys turning up in your doorknob unannounced at a later date. One last story about keys.

Never use a vendor from the job to re-key your locks. I'm familiar with an incident where this happened and the vendor ended up giving the builder an extra set of keys. They are obligated to the builder for future work. Don't let it happen to you!

EPILOGUE

Embracing this field guide will help you identify the PROS and understand and deal with the CONS and their games that may eventually surface during the building process.

Yet, informed individuals must remain vigilant to the twists and turns of what you have explored in this guide. Study the information contained in this resource, and you will be better prepared to protect your interest.

As you proceed through your building project gaining more experience and information, consult this guide for further insight and understanding. It will be helpful to revisit examples to further enhance your perspective as to what's really happening on your job site.

Knowledge is power! Make it yours! Remember it's your choice; **you get what you pay for, or you pay for what you get**.

Good luck!

Appendix A

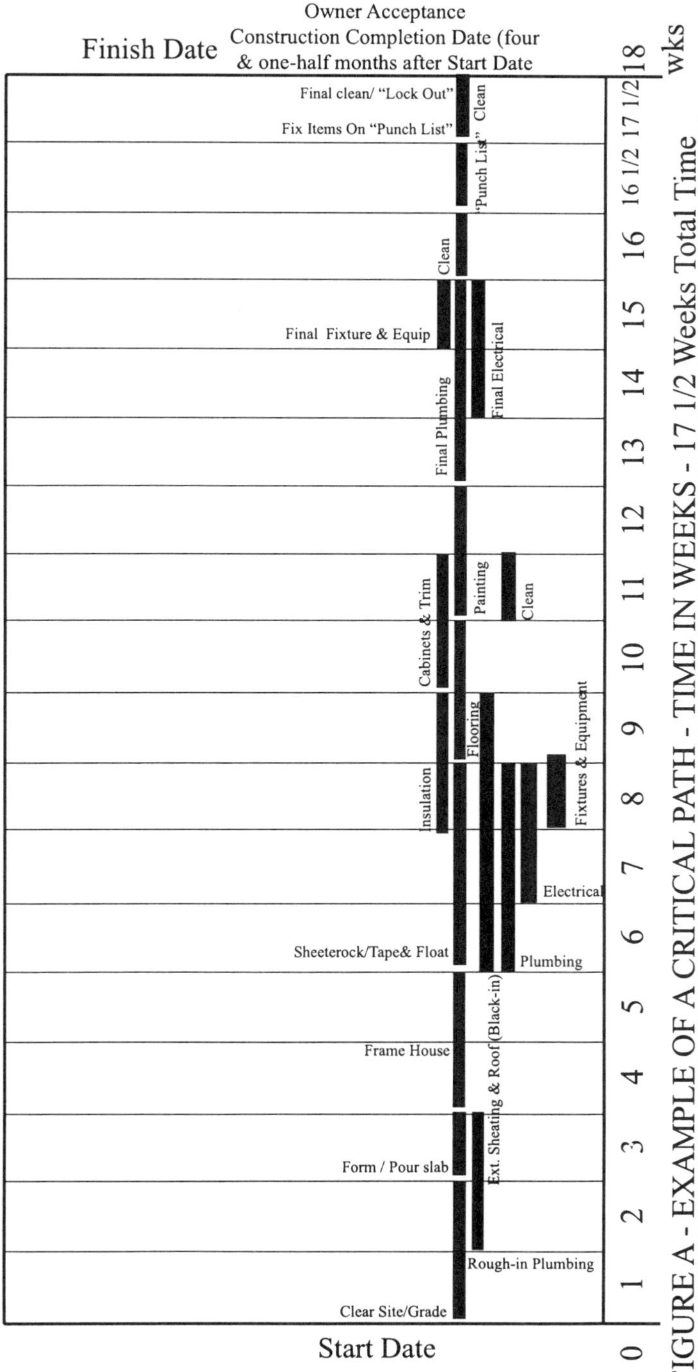

FIGURE A - EXAMPLE OF A CRITICAL PATH - TIME IN WEEKS - 17 1/2 Weeks Total Time

NOT TO BE USED FOR CONSTRUCTION

Reference

1.) A.I.A. (American Institute of Architects), 800-242-3837, www.aiaaccess.com

2.) AIBD (American Institute of Building Designers), 800-366-2423, www.aibd.com

3.) Faux-naïf House, L.L.C. to order books, purchase book outline and check list, communicate with the author, and to add your stories to the next update of Before You Build! www.fauxnaifhouse.com www.cloverchains.com

4.) Gypsum Association, 810 First Street, NE, #510, Washington, DC 20002, (202)289-5440, www.gypsum.org

5.) NAHB (National Association of Home Builders, 800-368-5242, www.nahb.com

6.) Pocket Size Carpenter's Helper (ISBN 0-937635-00- 6) R.S. Wood & Co., Belfast, Maine